THE AMERICAS

The global political map is undergoing a process of rapid change as former states disintegrate and new states emerge. Territorial change in the form of conflict over land and maritime boundaries is inevitable but the negotiation and management of these changes threaten world peace.

The Americas offers a wide-ranging and original interpretation of matters relating to territory, boundaries and societies in the American continent.

World Boundaries is a unique series embracing the theory and practice of boundary delimitation and management, boundary disputes and conflict resolution, and territorial change in the new world order. Each of the five volumes – *Middle East and North Africa, Eurasia, The Americas, Maritime Boundaries* and *Global Boundaries* – is clearly illustrated with maps and diagrams and contains regional case studies to support thematic chapters. This series will lead to a better understanding of the means available for the patient negotiation and peaceful management of international boundaries.

Pascal O. Girot is Associate Professor of Geography at the University of Costa Rica, San José.

WORLD BOUNDARIES SERIES
Edited by Gerald H. Blake
*Director of the International Boundaries Research Unit
at the University of Durham*

The titles in the series are:

GLOBAL BOUNDARIES
Edited by Clive H. Schofield

THE MIDDLE EAST and NORTH AFRICA
Edited by Clive H. Schofield and Richard N. Schofield

EURASIA
Edited by Carl Grundy-Warr

THE AMERICAS
Edited by Pascal Girot

MARITIME BOUNDARIES
Edited by Gerald H. Blake

THE AMERICAS

World Boundaries Volume 4

Edited by Pascal O. Girot

London and New York

CONTENTS

vi

FIGURES

FIGURES

TABLES

NOTES ON CONTRIBUTORS

Peter M. Slowe, Department of Geography, West Sussex Institute of Higher Education

Olivier Kramsch, Graduate School of Architecture and Urban Planning, University of California, Los Angeles

Cuauhtémoc Leon, Universidad Autonoma de Baja California

Marina Robles, Universidad Autonoma de Baja California

Allan Lavell, School of Architecture, University of Costa Rica

Jan de Vos, Centro de Investigaciones y Estudios Superiores (CIESAS), San Cristobal de Las Casas, Chipas, Mexico

Alfredo C. Dachary, Centro de Investigaciones de Quintana Roo, Chetumal, Mexico

Pascal O. Girot, Department of Geography, University of Costa Rica

Ronald Bruce St John, Caterpillar Far East

Bertha K. Becker, Department of Geography, Federal University of Rio de Janeiro

Susana Bandieri, Universidad Nacional del Comahue, Neuquén, Argentina

Monica Gangas Geisse, Instituto de Geografia, Pontificia Universidad Catolica de Chile

Hernan Santis Arenas, Instituto de Geografia, Pontificia Universidad Catolica de Chile

SERIES PREFACE

The International Boundaries Research Unit (IBRU) was founded at the University of Durham in January 1989, initially funded by the generosity of Archive Research Ltd of Farnham Common. The aims of the unit are the collection, analysis and documentation of information on international land and maritime boundaries to enhance the means available for the peaceful resolution of conflict and international transboundary cooperation. IBRU is currently creating a database on international boundaries with a major grant from the Leverhulme Trust. The unit publishes a quarterly *Boundary and Security Bulletin* and a series of *Boundary and Territory Briefings*.

IBRU's first international conference was held in Durham on September 14–17, 1989 under the title of 'International Boundaries and Boundary Conflict Resolution'. The 1989 conference proceedings were published by IBRU in 1990 under the title *International Boundaries and Boundary Conflict Resolution* edited by C.E.R. Grundy-Warr. The theme chosen for our second conference was 'International Boundaries: Fresh Perspectives'. The aim was to gather together international boundary specialists from a variety of disciplines and backgrounds to examine the rapidly changing political map of the world, new technical and methodological approaches to boundary delimitations, and fresh legal perspectives. Over 130 people attended the conference from 30 states. The papers presented comprise four of the five volumes in this series (Volumes 1–3 and Volume 5). Volume 4 largely comprises proceedings of the Second International Conference on Boundaries in Ibero-America held at San José, Costa Rica, 14–17 November, 1990. These papers, many of which have been translated from Spanish, seemed to complement the IBRU conference papers so well that it was decided to ask Dr Pascal Girot, who is coordinator of a major project on border regions in Central America based at CSUCA (The Confederation of Central American Universities), to edit them for inclusion in the series. Volume 4 is thus symbolic of the practical cooperation which IBRU is developing with a number of institutions overseas whose objectives are much the same as IBRU's. The titles in the World Boundaries series are:

Volume 1 *Global Boundaries*
Volume 2 *Middle East and North Africa*
Volume 3 *Eurasia*
Volume 4 *The Americas*
Volume 5 *Maritime Boundaries*

The papers presented at the 1991 IBRU conference in Durham were not specifically commissioned with a five-volume series in mind. The papers have been arranged in this way for the convenience of those who are most concerned with specific regions or themes in international boundary studies. Nevertheless, the editors wish to stress the importance of seeing the collection of papers as a whole. Together they demonstrate the ongoing importance of research into international boundaries on land and sea, how they are delimited, how they can be made to function peacefully, and perhaps above all how they change through time. If there is a single message from this impressive collection of papers it is perhaps that boundary and territorial changes are to be expected, and that there are many ways of managing these changes without resort to violence. Gatherings of specialists such as those at Durham in July 1991 and at San José in November 1990 can contribute much to our understanding of the means available for the peaceful management to international boundaries. We commend these volumes as being worthy of serious attention not just from the growing international community of border scholars, but from decision-makers who have the power to choose between patient negotiation and conflict over questions of territorial delimitation.

Gerald H. Blake
Director of IBRU
Durham, January 1993

PREFACE

Referring to the differences between Anglo-Saxon and Latin American culture, Jorge Mañach once wrote that 'in the duality of American cultures, the frontier constitutes . . . both one of risk and of privilege';[1] a boundary which simultaneously attracts and repels. Indeed, boundaries in America are frequently reduced to the often over-bearing frontier which separates the United States from the rest of Latin America. This widespread obsession with the Rio Grande as the single most important border in the Americas often leads scholars and politicians alike to overlook an immensely diverse and complex array of boundary and border configurations in the rest of the continent.

In response to this tendency, the present volume offers a synopsis of border situations across the American continent. In an effort to provide a significant geographical coverage, we have been forced to select works dealing with boundaries and border regions in representative areas of the continent. While the coverage is by no means exhaustive, excluding for instance important boundary situations such as those found in Colombia, Venezuela and Guyana, it does provide a good overview of border regions and related problems in the Americas. The order in which we have organized the contributions and the division into sections on North, Central and South America, also follows a somewhat classic North to South bias. This does not, however, imply geographic preferences of any sort.

Borderland studies are experiencing a significant boom in Latin America. Authors from a variety of disciplines, including History, Geography, Architecture and Planning, as well as Law and Political Science, are researching border phenomena and producing new knowledge on the conformation of boundaries in the Americas. This

1   Mañach, J. (1970) *Teoría de la Frontera*, San Juan: Editorial Universitaria, Universidad de Puerto Rico: 133 (Editor's translation).

wide variety of approaches is reflected in this volume, in which scholars from diverse academic perspectives offer original interpretations of matters relating to territory, boundaries and societies in the Americas. In spite of the wide-ranging approaches and geographic contexts of these contributions, common features can be identified. Peter Slowe's and Olivier Kramsch's chapters on the Quebec–US and Mexico–US borders respectively focus on the nature of economic flows across boundaries, and policy responses to changing situations. In an innovative approach, Cuauhtémoc Leon and Marina Robles deal with the interaction between California and Baja California from an ecological perspective. In an interesting contribution on Central American boundaries, Allan Lavell proposes a typology of border regions and a future research agenda. Many of the problems identified by Lavell are subsequently treated by authors such as Jan de Vos on Chiapas, Mexico's border state with Central America, and Alfredo Dachary on the Mexico–Belize boundary. Pascal Girot's chapters focuses on the particular geopolitical configuration of the Central American Isthmus, and the influence of the interoceanic canal projects on the boundary configuration between Costa Rica and Nicaragua.

In the case of South America, four chapters provide a contrasting overview of the problems related to territory, frontiers and boundaries in the region. Ronald Bruce St John offers a succinct overview of the Peru–Ecuador dispute, perhaps one of the region's most volatile. Brazil constitutes a territory of sub-continental proportions, and deserves to be treated from a distinct perspective. In this sense, Bertha Becker provides an incisive vision of the complex interaction between technology, geopolitics and frontiers in Brazil. Her innovative approach is bound to open the way to future research on the role of the State and technology in territorial politics. Susana Bandieri presents a detailed historical account of a particular brand of transboundary interaction, in the border region of Neuquén between Argentina and Chile. Finally, Monica Gangas Geisse and Hernan Santis Arenas provide an interpretation from the Chilean perspective of Bolivia's maritime aspirations.

As the reader will appreciate, the wide-ranging topics and regions treated in this volume constitute an original contribution to the study of boundaries in the American continent. It is our sincere hope that it will provide scholars, students and policy makers alike with essential reading on the subject as well as open the way for future research on boundaries and border regions in the Americas.

Pascal O. Girot

Santa Elena de Zurquí, July 1992

ACKNOWLEDGEMENTS

Much of the initial work on these proceedings was undertaken by IBRU's executive officer Carl Grundy-Warr before his appointment to the National University of Singapore early in 1992. It has taken a team of editors to complete the task he began so well. Elizabeth Pearson and Margaret Bell assisted in the preparation of the manuscripts for several of these volumes, and we acknowledge their considerable contribution. In addition many people assisted with the organization of the 1991 conference, especially Carl Grundy-Warr, Greg Englefield, Clive Schofield, Ewan Anderson, William Hildesley, Michael Ridge, Chng Kin Noi and Yongqiang Zong. Their hard work is gratefully acknowledged. We are most grateful to Tristan Palmer and his colleagues at Routledge for their patience and assistance in publishing these proceedings, and to Arthur Corner and his colleagues in the Cartography Unit, Department of Geography, University of Durham for redrawing most of the maps. We are particularly grateful to Professor John Dewdney of Durham University for assisting with the editing of the final draft of this volume.

GLOSSARY OF SPANISH TERMS

Alcalde	Mayor
Alcaldías	Municipal or town authorities
Audiencia	Territorial and administrative division of the Spanish colonial possessions in South America, smaller than the vice-kingdom and larger than the capitania
Cabildos	Town or popular councils, or members of these councils
Cedula	A royal decree by which land grants and other administrative matters were established under the Spanish Crown
Encomienda	Land grants conferred by the Spanish Crown to Spanish settlers in the Americas, including land and population liable for tribute
Exposicion	In diplomatic and foreign affairs, an official presentation of a country's position concerning a political or territorial issue
Intendencias	A military and administrative term defining the territory controlled by an *Intendente*
Mandamiento	Writs by the Spanish Crown which defined the property and land-ownership rights of communities
Marquisate	The territory controlled by a Marquis

Part I

BOUNDARIES IN NORTH AMERICA

1

THE GEOGRAPHY OF BORDERLANDS

The case of the Quebec–US borderlands

Peter M. Slowe

INTRODUCTION

Borderlands on either side of undefended and easily crossed boundaries are now more common than ever. Economic pacts and political change in Europe, North America and Africa have changed dramatically the character of the boundaries and borderlands of three continents. The aim of this paper is to examine the impact of the increasingly open boundary between Quebec and the United States on the borderlands on the Quebec side, and thus to throw some light on this important aspect of regional geography.

As Prescott (1987: 159–73) has pointed out, boundaries have a significant impact on the economy, culture and environment of the borderlands, but there has been a neglect of their study in many areas. There are exceptions as far as the United States is concerned; there is House's (1982) monograph on the Rio Grande border between the United States and Mexico and recent studies of the general nature of the Canada–United States boundary produced by the University of Maine Borderlands Project, notably McKinsey and Konrad (1989) and Blaise (1990). But for the most part the extensive scholarship on the relations between Quebec and the United States takes no account of the impact of the boundary on the borderlands. Some recent work on the Canada–United States Free Trade Agreement has stressed the effects of the Agreement on the economy of the United States side of the boundary (Gandhi 1990; Lessler 1990), but until recently only a few specialized environmental studies have concerned themselves with the Canadian side (for example Carroll 1983: 104–13), apart from the survey of the Quebec–Maine border by Sanguin (1974), which also discussed some of the incidental problems for local people arising from alignment anomalies.

Otherwise the focus of the literature on relations between Quebec and the United States has been above all on the economic opportunities presented by Quebec's increasingly prosperous economy and on the threat to Quebec culture from Americanization following the era of the Quiet Revolution. Hero and Balthazar's (1988) major study would be an example of these twin emphases, while the work of Proulx and Shipman (1986) and the Ministère du Commerce Extérieur et du Développement Technologique (1988) on preparations for, and the likely impact of, the Free Trade Agreement exemplify the commercial aspect of the work. The extensive literature on cultural relations is couched in general terms and is not concerned with the borderlands, for example, Chartier (1984) on the positive aspects of cross-border contacts and Jones (1984) on the spectre of Americanization. While all this is essential background for a study of the Quebec borderlands it is indeed only background.

The boundary between Quebec and the United States is patrolled but not defended. Its openness encourages economic, environmental and cultural links; but the very fact that it is a political barrier, like all international boundaries, inevitably reduces both the informal exchange and the administrative cooperation, whether voluntary or enforced, which characterize the relations between neighbouring places within states.

The main impact of the boundary is inevitably on the immediate vicinity, or borderlands. In the case of Quebec these are defined as the municipalities of which part lies within twenty kilometres of the boundary, accounting for a total area of some 20,000 square kilometres and a total population of some quarter of a million.

THE QUEBEC–UNITED STATES BOUNDARY

The boundary between Quebec and the States of Vermont and New York was originally demarcated by the British between 1771 and 1774, and that between Quebec and New Hampshire and Quebec and Maine was settled in the 1842 Webster–Ashburton Treaty after at least two decades of disagreement (Nicholson 1954: 93–4). It has never been a militarily defended boundary, although there continue to be customs and immigration patrols to prevent the illegal movement of goods and people. Crossing the boundary is easy, especially for local people who have to pay only cursory attention to customs and immigration officials.

During the period of American prohibition, the smuggling of whisky from Canada into the United States was one celebrated cause of increased patrols along the boundary (Comité Organisateur 1989: 22). Currently, the main concern of customs controls has been to tax goods being taken into Canada which do not come under the rules of the Free Trade Agreement or other similar agreements, and the main concern of immigration controls has been to prevent the entry of illegal immigrants, especially originating from Asia, into the United States.

The United States borderlands in Maine and New Hampshire are for the most part uninhabited, but the towns and villages of Northern Vermont and Northern New York have close day-to-day ties with their neighbours on the other side of the boundary. These ties, which have a considerable impact on the borderlands, are both private and official, and it is official ties which have been enhanced by a series of agreements between Quebec and all four states. The first agreement of any significance, as far as its impact on the Quebec borderlands was concerned, was the agreement with Maine in 1972, covering education, culture, youth exchange and general economic cooperation including matters to do with forestry and the movement of labour (Gouvernement du Québec 1984: 29–30). More than a decade later, in the late-1980s, there were two crucial environmental agreements: with New York and Vermont over Lake Champlain (Gouvernement du Québec 1988) and with New Hampshire over Lake Mephremagog (Gouvernement du Québec 1989a). Important cultural agreements were also made at this time with both New Hampshire (Gouvernement du Québec 1989b) and Vermont (Gouvernement du Québec 1989c).

The federal constitutions of both Canada and the United States give sufficient autonomy to provinces and states to enter into international agreements covering most cultural and environmental matters, but covering economic matters only in so far as they have no direct effect on trade. Despite the well-known pressure for various amounts of additional autonomy for Quebec within Canada, Canadian federal law on international trade, immigration (with certain minor variations) and the administration of the boundary remains the same in Quebec as the rest of Canada.

Unlike the provincial or state authorities of Quebec, Maine, Vermont, New Hampshire and New York, however, local authorities on either side of the boundary do not have the legal right to enter into formal international agreements, but only into loose and unenforceable understandings. Yet, as this paper shows, it is at this local level that the boundary has its greatest impact on the people and the landscape and that the north–south axis between Quebec and the New England states has perhaps its profoundest imprint.

The borderlands economy

The Free Trade Agreement between Canada and the United States has had a more important impact on the Quebec borderlands through the way it has been popularly perceived than through its actual terms. Eighty per cent of exports from Canada to the United States and 65 per cent of imports by Canada from the United States were freely traded before the Agreement started to be implemented in January 1989, and only 15 per cent of the remainder were removed in 1989, with a further 35 per cent planned to be removed by 1994. Furthermore, the agreement does not ease restrictions on the movement of labour and does not in any sense create a European-style common market; it is first and foremost an international trade agreement. So, even though the United States accounts for nearly 80 per cent of Quebec's exports, the impact of the agreement on the borderland economy is unlikely to be great.

The biggest effect of the agreement as it is perceived and one of the main characteristics of the borderland economy is the continuing rundown of the retail sector in the Quebec borderlands and its equivalent growth on the United States side. A sample of twenty retailers in the Quebec borderlands were interviewed in August 1990 in Coaticook, Hemmingford, Huntingdon, Magog, Philipsburg, Pohénégamook, Rock Island, St Georges and Stanstead. They estimated that some 25 per cent of dairy products, beef, alcoholic beverages, coffee, tobacco, electronic goods and petrol consumed locally were purchased in the United States. There had been an increase of about 50 per cent over the two years from 1988 to 1990 with a related loss of jobs. These figures for settlements near the boundary were rather higher than comparable figures produced for British Columbia (Hamilton 1990), but the Quebec estimate was for areas without major cities very close to the boundary on the Canadian side, and it also takes more smuggled goods into account than the British Columbian survey, which was mainly concerned with dairy products.

The principal reasons for the cross-border shopping phenomenon as a whole include agricultural goods supply management by marketing boards in Canada, higher wage costs in Canada, differential taxation on petrol, and the greater variety of goods available in the United States (in part because the French language rule for packaging does not apply outside Quebec). The reasons for the increases associated with the start of the implementation of the Free Trade Agreement were the intense marketing campaign by United States borderland retailers, who cater specifically for Quebeckers' tastes, using the French language and quoting prices in Canadian currency, to exploit a popular misconception about cheaper and easier cross-border shopping; shopping malls in Burlington, Jackman, Newport and Plattsburgh received substantial investment for expansion and marketing from both US and Canadian investors, and they rely on Quebeckers for between one-third and two-thirds of their sales. 'A lot of people think that the Free Trade Agreement means they can go to the other side and buy whatever they want ...', one Canadian customs officer complained in January 1989 (Scowen 1989: i), but 'the Agreement doesn't really mean much; whisky and rum are duty-free now but there's still the sales tax and the excise duty'. But at the same time the stores in Newport just across the boundary were planning for a 20 per cent increase in trade just to cater for a successful marketing campaign exploiting nothing more than a vague feeling that buying cheaper goods in the United States would be easier after the agreement came into force. It is clear that, although the initial impact on what most Quebeckers actually wanted to buy in the United States was very slight, the Free Trade Agreement was an effective instrument for marketing the cross-border shopping idea. It has been a golden opportunity to proclaim the benefits of cross-border shopping, already well known to 30 per cent of Quebeckers, to an even wider audience.

This increase in cross-border shopping may be expected to have mixed effects on the borderlands economy. First, retail stores will be faced with the need to reduce their operating costs or with the inevitability of losing more sales and employment. Labour costs are the highest component of operating costs, especially in the chain stores, so unions will be weakened and pay scales are likely to be gradually reduced.

Second, supply management will be hard to maintain, although farm support programmes, including income stabilization and marketing boards, were allowed to continue under the Free Trade Agreement. Attempts to enforce stricter adherence to border controls to stop people buying cheaper food a few kilometres away across the boundary have the potential to make the whole supply-management system politically untenable. Grubel (1990: 13–14) explained the political dilemma with the opposite example of *de facto* free trade in processed foods like soups, frozen dinners, yoghurt and ice-cream which threatened to drive out of business Canadian processors who would still have to buy from the artificially high-cost Canadian marketing boards. The suggested solution was to increase the retail prices of milk, eggs, chickens and so on, so that prices to processors could be reduced. However, by solving the problem of processed food price differentials, the problem of fresh food would increase. Grubel's conclusion was that it is inevitable that supply management will wither away; if it does so, one of the main reasons for regular cross-border shopping will disappear and the borderlands retail sector will revive.

The United States borderlands have benefited from substantial Canadian investment, in part by firms wishing to take advantage of cross-border shopping but mainly by firms wishing to retain close links with parent companies in Quebec while expanding into the United States market. It was thought that the Free Trade Agreement, with the reduction of tariffs and the easing of business travel, would bring about a transfer of some investment from the United States borderlands to the Quebec borderlands, because they would have virtually the same locational advantages for Canadian investors without having to deal with two administrative systems. But a survey of Canadian investment in northern New York State by Gandhi (1990) showed that there would be no transfer and that the main reason for location in the United States was not to overcome tariff barriers but to overcome non-tariff barriers to the United States market, especially variations of the 'Buy American' slogan. Investment in warehousing and marketing outlets in the United States borderlands is still on the increase (Bryan 1990; Nadeau 1990), while Quebec's higher wage costs (estimated at some 19 per cent higher in retail) and French language policy, which means, effectively, that everything has to be done in two languages, have been additional reasons for investment just over the international boundary.

A vital part of the economy of the Quebec borderlands is the forest industry which is responsible for about 15 per cent of all employment (with thirty-five sawmills and five pulp mills); in the more sparsely populated area of the border with Maine it accounts for as much as 40 per cent (Association des Industries Forestières du Québec 1990: 4–5). As Don (1989: 9) points out, forestry issues delayed the Free Trade Agreement but in the end were little affected by it. Canadian subsidy schemes for softwood lumber were left nearly intact and wood pulp and newsprint paper were generally free of duty anyway.

A separate agreement, however, reflected the pressure on the US Government resulting from cheap imports of British Columbian softwood lumber in the far west. To obtain other agreements more easily and to avoid a risk of new US duties, the Canadian Government agreed to additional taxation on forest industries as a whole (while still allowing certain local subsidies) and to restrictions on the export of unprocessed or partially processed lumber. This was a federal decision over which Quebec had little influence. The previously prosperous softwood lumber sawmills in Quebec suffered immediately from this new regime, especially those in borderland locations. The sawmills along Quebec's boundary with Maine, many of them actually owned by US companies, get almost all their timber from across the boundary in the unpopulated North Maine Woods and re-export it semi-processed. The tax (which may now be abolished) and the export restrictions have increased the number of sawmills at risk on Quebec's boundary with Maine to 50 per cent.

To compound the problems of the border operators, increasing tracts of forest in Maine are now owned by landlords giving priority to the long-term development of hunting aud outdoor activities. Stumpage charges, the amount paid for each tree felled, in Maine have risen accordingly. At the same time, a new law in Quebec aimed at environmental protection has restricted access to the forests on the Quebec side of the boundary to those firms exploiting lumber the previous year; no new entries were permitted so firms using timber from Maine had no access to competitive sources if charges increased. The new law also resulted in an increase in material costs within six months of between 10 and 15 per cent for the pulp mills in the borderlands between Vermont and New York. Previously they used cheap imported timber from across the border, but for a year now they have used Quebec timber just to ensure their future allocation, a costly and uncompetitive operation.

Other sectors of the borderlands economy are not affected by the border as such. In these cases, the impact of the Free Trade Agreement and other boundary characteristics is no different from that on their counterparts elsewhere in Quebec. The textile industry, for example, is threatened by cheaper US imports which are partly made possible through the use of cheap labour in Mexico or Haiti. On the other hand, the extraction and export of granite and tombstones on Quebec's border with Vermont, which supports over three hundred employees in two villages, will benefit from increased possibilities in penetrating the US market and from the possibility of increased binational lobbying for protection from Third World competition (Scowen 1988: 1 and 11).

Finally, the tourist industry is important in most of the Quebec borderlands. At the local level, cheaper food and labour means cheaper restaurants, and typically half of the eating-out advertised in the borderlands is across the boundary in the USA. The main movement in the opposite direction is of youths between the ages of eighteen and twenty-one who can be served alcohol in Quebec but not in the border states. On Quebec's border with Maine, the villages of Eastcourt and Pohénégamook benefit from boundary crossing points giving access to forest activities in the uninhabited North Maine Woods for which they can sell tourist facilities and equipment and deal with much of the administration. General tourism in the borderlands depends, as everywhere else in Canada, on the prevailing state of the economy in Canada and the United States, cultural and demographic changes, and the relative value of the Canadian and American dollars (Bellerose 1988).

The borderlands environment

Most environmental matters, such as airborne pollution, acid rain and wildlife and habitat conservation are inevitably cross-border issues, but for the most part are not particularly or specifically relevant to places near the boundary. It is around the border lakes, especially Lakes Champlain and Mephremagog, that conflicts of interest directly impinging on the borderlands have to be worked out. The resulting environmental diplomacy has a significant impact on the geography of the Quebec borderlands.

The most important water boundary between Quebec and the United States is Lake Champlain, a large and scenic lake surrounded by Vermont's best agricultural land and several of its main population centres on its east shore, a less important corner of New York on its west shore, and Quebec in the north where the Richelieu River drains. As a large boundary lake, it is covered by the Boundary Waters Treaty 1909 (United States of America 1909) which established the International Joint Commission for resolving any international disputes.

The international dispute arising from Lake Champlain really came to light in the early-1970s when the lake level was high and the Richelieu River consequently flooded homes recently built in its floodplain. These were mostly Montrealers' second homes, originally unpopular with many of the locals, mainly because they pushed up land prices. The Richelieu had in fact flooded twenty years before and a dam to protect farmland had been affected. Quebeckers demanded a dam to prevent flooding, but this would have resulted in a massive loss of wetlands rich in wildlife just at a time when environmental conservation was becoming politically important in the United States. In Quebec, the issue was soon seen as a question of American ducks against Quebeckers' homes, whereas in Vermont in particular the issue was seen as the heritage of the people of Vermont against sensible floodplain

10

zoning and insurance (Carroll 1983: 105).

The International Joint Commission responded to the problem by setting up two boards to study the problem: first an engineering board, which concluded that the waters could be regulated but at an indeterminate cost to the environment; a further study was then set up which concluded that a mixture of flood control and floodplain zoning was needed. The balance of the report was, increasingly unfashionably, insensitive to the wetlands issue. In particular, the method used to evaluate the economic value of wetlands as against floodplain housing and the possible use of dried-out wetlands tended to undervalue the importance of the lake's recreational and tourist potential. Also, in suggesting the use of dyking, culverts and pumping to create more wetlands (even though it also concluded that no remedial measures were actually necessary) the Board showed a surprising lack of ecological understanding that flawed the report as a whole (International Champlain–Richelieu Board 1978: 5–8). The resulting outcry by environmentalists from everywhere except the Quebec borderlands, and in particular the lobbying by the Lake Champlain Committee which was active right through the 1970s and still is today, resulted in nothing being done. The International Joint Commission (1981) finally failed to make any recommendation and it is now likely that nothing significant will be done to control flooding. Doing nothing has so far taken the best part of twenty years, such is the slow pace of resolving international disputes of this sort.

The French-speaking Quebeckers who bought second homes in the ill-fated Richelieu floodplain sometimes compared the apparent lack of resolve in their case by the Federal Canadian Government with the case of Lake Mephremagog further east along the boundary with Vermont, whose shoreline was surrounded mainly by the homes of the Anglophones from that district (Carrol 1983: 110). This accusation ignores the different administration of Lake Mephremagog which is crossed by the boundary transversely and is therefore beyond the remit of the International Joint Commission under the 1909 Treaty.

Lake Mephremagog has had an effective barrage since 1882 and, perhaps for that reason, has never had the administrative problems of the much larger Lake Champlain. The shores of Lake Mephremagog are one-fifth in Vermont and four-fifths in Quebec with Newport (Vermont) at the southern end and Magog (Quebec) at the northern tip. Technically, the responsibility for controlling the limited amount of navigation remains with the federal governments, although the lake bed is the responsibility of Quebec and Vermont, and all the powers at the lake edge over planning, including anti-pollution measures, lie with the municipalities in Quebec and the cantons in Vermont. A powerful technical and advisory organization was set up by the two federal governments headed by an engineer from each side primarily concerned with lake levels and flooding, in this case a concern on both sides of the boundary; this was the Conseil International de Contrôle (Conseil International de Contrôle du Lac Mephrémagog 1983). While the main concern was the lake level and not pollution the Conseil was effective, but it never had sufficient authority to pronounce on the environmental issues with which increasingly it had to deal: for example, in clashes with cantons over the design of marinas and the use of septic tanks, the Conseil was powerless. Certain informal agreements were made with Mephremagog Conservation Incorporated which the Conseil had played a crucial part in setting up in 1967, which helped to clean up the lake. By late 1987, however, there were increasingly serious problems of overnutrification and pollution by both sewage and fertilizers. The Municipalité Régionale de Comté de Mephrémagog (the regional grouping of municipalities on the Quebec side), with the informal approval of the Vermont side, approached the Quebec Government to investigate the administration of the lake with its environmental problems in mind. The result was the Environmental Cooperation Agreement on Managing the Waters of Lake Mephremagog (Gouvernement du Québec 1989a) between Quebec and Vermont, aimed at investigating the administrative problems. The resulting Quebec–Vermont Working Group on Managing Lake Mephremagog and its Environment held its first meetings in July and September 1990 and, although there were only six members and no representation for voluntary groups, early indications were promising. The appointed members were high-ranking officials with substantial support staff and there was every sign of the group interpreting its remit widely and dealing directly with the liaison of environmental policy. In particular, coordination between local authority plans for the lake on the Quebec side (Municipalité Régionale du Comtede de Mephrémagog 1988) and the Vermont side (Department of Environmental Conservation 1990; Northeastern Vermont Development Association 1990) looks like being made possible by this centralized and powerful group, albeit with very limited formal powers (Quebec–Vermont Working Group on Managing Lake Mephremagog and its Environment 1990).

The conservationists at Mephremagog, like the second-home owners in the Richelieu floodplain, are nevertheless the victims of borderland bureaucracy, where a complex international dimension, which in practice greatly reduces the effectiveness of voluntary organizations, is added to already difficult environmental problems. The frustrations of this borderland situation showed up in angry accusations between language groups in the case of Lake Champlain and in frustration, inaction and splits in the conservation movement between constitution-alists and activists not prepared to wait (Scowen 1990) at Lake Mephremagog.

The borderlands culture

Quebec's Quiet Revolution gave Quebeckers the confidence to escape the territory of Quebec. In Feldman's words (Duchacek 1986: 276–7): 'Quebec has enlarged and polished its international profile'; Quebec nationalism had been constructively redirected (Slowe 1990: 74–7). Quebec entered into cultural agreements far and wide during the 1970s and 1980s. In the case of its American neighbours, Quebec signed an agreement with New Hampshire in 1989 (Gouvernement du Québec 1989b) and cultural exchanges have already begun, especially language immersion programmes for business people and schoolchildren (Ville de St Georges 1990). In the case of Vermont, the Commission Mixte is already well advanced with its cultural work on the boundary itself (Commission Mixte Québec–Vermont 1990), notably having reinstated the Haskell Opera House which is half in Quebec and half in Vermont.

This is small-scale when compared to the cultural rewards of regular interchange typical of all open boundaries between two cultures. The boundary has a clearly marked and patrolled physical presence but it does not always mark a clear cultural or even linguistic divide. In New Hampshire and Maine there are Francophone Americans, and in the southern townships on Quebec's border with Vermont and New York there are many Anglophone communities with the right to have bilingual signs and able to take advantage easily of American cinemas and other entertainment. The people in the border towns live on each other's doorsteps, playing sport together, shopping in each other's shops, eating in each other's restaurants, and drinking in each other's bars. Intermarriage and dual nationality are common, the latter especially on the Quebec side because, before state welfare started in Quebec, the best and nearest maternity hospitals were frequently over the boundary. For the people of the borderlands, regular exchanges destroy the false and negative perceptions which can otherwise emerge; it is notable, for example, that the disputants in the Champlain–Richelieu case are not primarily the border people, but as at Lake

Mephremagog, it is a case of the people on both sides of the boundary against their respective provincial and state bureaucracies.

Perhaps the strangest borderlands phenomenon in Quebec occurs at Akwesasne on the extreme west of the boundary. This is a Mohawk Indian reserve which spreads out from their tri-junction into New York, Ontario and Quebec. Like all such reserves, Akwesasne is independent of the states and provinces and subject to federal jurisdiction loosely administered through the Six Nations Iroquois Confederacy. It has built its economy on gambling (illegal in New York, Ontario and Quebec but legal and approved by referendum in the reserve) and on smuggling, especially cigarettes, across the boundary which the Mohawks themselves generally patrol. The organizers of the gambling and smuggling, the self-styled 'Warriors', have linked militant demands for better treatment for Indians (Maxwell 1990) to a heavily armed protection racket associated with illegal activities. On the New York side of the boundary, the state police moved in May 1990 to disarm the Warriors and urged the Canadian authorities to do the same (Heinrich 1990), but in Canada long-term diplomacy and overdue economic concessions seem more likely in the case of Akwesasne, although force has been used elsewhere.

It is ironic that the people to whom the boundary means nothing historically or culturally have come to depend on it. It enhances their income through smuggling. It enables them to play one authority off against another and to seek refuge on one side or the other. They can use violence on one side of the boundary and claim their lost rights through court action on the other. The boundary itself has become one of their weapons.

CONCLUSION

In the case of the Quebec borderlands with the United States, there are a number of geographical features explained by the presence of the boundary. The most important economically are the decline of the retail sector and the threat to the forestry sector, neither significantly affected by the recent Canada–US Free Trade Agreement. The administration of the border lakes, Champlain and Mephremagog, has been made cumbersome by the boundary which runs through them; on this open boundary, inertia has been the hallmark of environmental policy in contrast to more vigorous activity by provincial, state and federal governments away from the boundary. Culturally, the daily cross-boundary exchange gives a special character to the border region.

14

The phenomenon of borderlands next to open international boundaries deserves further attention by geographers. The economic, environmental and cultural geography of borderlands are all crucially influenced by the special locational characteristic of simply being near a boundary.

BIBLIOGRAPHY

Association des Industries Forestières du Québec (1990) *L.A.I.F.Q.*, Quebec: Association des Industries Forestières du Québec.

Bellerose, P. (1988) *Le libre-échange et le champ recréotouristique: un secteur gagnant*, Quebec: Centre d'Etude du Tourisme.

Blaise, C. (1990) *The Border as Fiction*, Orono: University of Maine Borderlands Project, Borderlands Monograph Series, No. 5.

Bryan, J. (1990) 'Entrepreneurs in U.S. border towns look northward', *The Gazette* (Montreal), May 15, 1990: 13.

Carroll, J.E. (1983) *Environmental Diplomacy: An Examination and a perspective of Canadian–U.S. Transboundary Environmental Relations*, Ann Arbor: University of Michigan Press.

Chartier, A.B. (1984) 'Franco-Americans in Quebec: linkages and potential', in Hero, A.O. and Daneau, M. (eds) *Problems and Opportunities in U.S.– Quebec Relations*, Boulder: Westview.

Comité Organisateur (1989) *Rivière Bleue: 75 ans d'histoire 1914–1989*, Rivière Bleue: Comité Organisateur.

Commission Mixte Québec–Vermont (1990) 'Joint meeting of the Vermont/ Quebec Commission, Friday, May 4, 1990', Radisson Hotel, Burlington, Vermont, unpublished minutes.

Conseil International de Contrôle du Lac Mephrémagog (1983) *Histoire du Conseil International de Contrôle du Lac Mephrémagog*, Quebec: Direction Générale des Eaux Intérieures.

Department of Environmental Conservation (1990) *Shoreland Zoning Options for Towns* (revised May 1990), Waterbury: Department of Environmental Conservation (Lake Protection Program).

Don, J.P. (1989) 'The Canada–United States Free Trade Agreement: regional impacts and policy responses: the case of New York State', paper given to the Tenth Biennial Meeting of the Association of Canadian Studies in the United States, November 17–20, 1989.

Duchacek, I. (1986) *The Territorial Dimension of Politics: Within, Among and Across Nations*, Boulder: Westview.

—— (1989) 'Le bilan d'une transition: quand la prudence frise l'imprudence', *Le Papetier*, 25 (March 1989): 1.

Gandhi, P. (1990) 'Free Trade Agreement and Canadian investment in northern New York', paper given to the conference 'The Canada–United States Free Trade Area: A year later', Cornier Center of International Economics, Bishop's University, Lennoxville, Quebec, March 21–22, 1990.

Gouvernement du Québec (1984) 'Communiqué conjoint: visite du Gouverneur du Maine, M. Kenneth J. Curtis, 6 Mai 1972', in Gouvernement du Québec

Recueil des Ententes Internationales du Québec, Quebec: Gouvernement du Québec.

Gouvernement du Québec (1988) *Entente intergouvernmentale sur la coopération en matière d'environnement relativement à la gestion du Lac Champlain entre l'Etat de New York et l'Etat du Vermont avec la participation du Gouvernement du Québec*, Quebec: Gouvernement du Québec.

Gouvernement du Québec (1989a) *Entente De Coopération en matière d'environnement relativement è la gestion des eaux du Lac Mephrémagog entre le Gouvernement du Québec et le Gouvernement de l'Etat du Vermont*, Quebec: Gouvernement du Québec.

Gouvernement du Québec (1989b) *Agreement On Cultural Cooperation Between the Government of the State of New Hampshire and the Gouvernment of Quebec*, Quebec: Gouvernement du Québec.

Gouvernement du Québec (1989c) *Memorandum Of Understanding (between le Gouvernement du Québec and the Government of the State of Vermont)*, Quebec: Gouvernement du Québec.

Grubel, H.G. (1990) *'Border-trade, free trade and interest groups'*, Simon Fraser University, Department of Economics, Economics Discussion Paper 90.

Hamilton, G. (1990) 'Suit seeks to enforce duty', *Vancouver Sun*, February 6, 1990: 1.

Heinrich, J. (1990) 'Ottawa urged to cripple warrior society', *The Gazette* (Montreal), July 25, 1990: 1.

Hero, A.O. and Balthazar L. (eds) (1988) *Contemporary Quebec and the United States 1960–1985*, Boston: University Press of America.

House, J.W. (1982) *Frontier on the Rio Grande: A Political Geography of Development and Social Deprivation*, Oxford: Clarendon.

International Champlain–Richelieu Board (1978) *Regulation of Lake Champlain and the Upper Richelieu River: Supplemental Report to the International Joint Commission*, Hull/Albany: International Champlain–Richelieu Board.

Jones, E.A. (1984) 'Le Spectre d'Américanisation', in Savary, C. (ed.) *Les Rapports Culturels entre le Québec et les Etats Unis*, Quebec: Institut Québecois de Recherche sur la Culture.

Lessler, A. (1990) 'The reaction of businesses, industries and residents of Clinton County, N.Y. to the FTA', paper given to the conference 'The Canada–United States Free Trade Area: a year later', Cornier Center of International Economics, Bishop's University, Lennoxville, Quebec, March 21–22, 1990.

McKinsey, L. and Konrad, V. (1989) *Borderland Reflections: The United States and Canada*, Orono: University of Maine Borderlands Project, Borderlands Monograph Series, No. 1.

Maxwell, L. (1990) 'Let's talk peace', *People's Voice*, August 3, 1990: 5.

Ministère du Commerce Extérieur et du Développement Technologique (1988) *The Canada–United States Free Trade Agreement: A Quebec Viewpoint*, Quebec: Ministère du Commerce Extérieur et du Développement Technologique.

Municipalité Régionale Du Comté De Mephrémagog (1988) *Schéma d'Aménagement*, Magog: Municipalité Régionale du Comté de Mephrémagog.

16

Nadeau, J.B. (1990) 'Le Vermont: votre camp de base pour l'assaut du marché Americain', *Petites et Moyennes Entreprises*, July/August 1990: 27–34.

Nicholson, N.L. (1954) *The Boundaries of Canada, its Provinces and Territories,* Ottawa: Department of Mines and Technical Surveys, Geographical Branch, Memoir 2.

Northeastern Vermont Development Association (1990) 'Memorandum: shoreland zoning of VT towns, 19 September, 1990', unpublished.

Prescott, J.R.V. (1987) *Political Frontiers and Boundaries,* London: Allen and Unwin.

Proulx, P. and Shipman W.D. (1986) 'Trade relations among Quebec, the Atlantic provinces and New England', in Shipman, W.D. (ed.) *Trade and Investment Across the Northeast Boundary: Quebec, The Atlantic Provinces and New England,* Montreal: Institute for Research on Public Policy.

Quebec–Vermont Working Group on Managing Lake Mephremagog and its Environment (1990) 'Meeting of the working Group held on July 16 1990', unpublished minutes.

Sanguin, A.L. (1974) 'La frontière Québec–Maine: quelques aspects limnologiques et socio-économiques', *Cahiers de Géographie du Québec,* 18 (No. 43): 159–85.

Scowen, P. (1988) 'Free trade: will it be good for local businesses?', *Stanstead Journal,* November 9, 1988: 1 and 11.

—— (1989) 'Local customs agents set to deal with free trade and uninformed public', *Stanstead Journal,* January 4, 1989: 1.

—— (1990) 'New group created after MCI fallout', *Stanstead Journal,* August 1, 1990: 1.

Slowe, P.M. (1990) 'Nationhood and statehood in Canada', in Chisholm, M.D.I. and Smith, D.M. (eds) *Shared Space: Divided Space: Essays on Conflict and Territorial Organization,* London: Unwin Hyman.

United States of America (1909) *Treaty Relating to Boundary Waters and Questions arising along the Boundary between the United States and Canada, January 11, 1909, Statute 36/2448,* Washington, DC: Congress of the United States of America.

Ville de St Georges (1990) 'Projets de développement majeurs: Franco-Américains de Nouvelle-Angleterre', unpublished memorandum.

2

TRANSBORDER REGIONAL PLANNING IN THE CONTEXT OF BINATIONAL ECONOMIC INTEGRATION

The case of a new border crossing between Mexico and the United States

Olivier Kramsch

INTRODUCTION

The environmental impact of accelerating economic integration between Mexico and the United States is nowhere more apparent than at the border separating the two nations. As the flow of capital, labour and goods passing from one country to the other has intensified, the US–Mexico border region has become a microcosm of an increasingly articulated binational economy. The negative consequences of increased tourist and commercial traffic passing through international border-crossing checkpoints has been acutely perceived on either side of the political dividing line, constituting a 'transfrontier externality' (Herzog 1990).

This chapter seeks to establish a methodological framework for transborder regional planning between the United States and Mexico by analysing the feasibility of constructing a new international border crossing near the Pacific Ocean between the cities of Tijuana, Baja California and San Diego, California. A major premise underlying the study is that both cities, though separated by a boundary which defines the political extremities of two distinct national subregions, form part of an interdependent urban ecosystem which transcends the border proper.

The feasibility study for a new border crossing was guided by the following criteria:

1   The potential demand for a new border-crossing facility within the

overall context of urban growth in Tijuana and San Diego, which in turn is determined by:

(a) processes of economic restructuring in each city and its resultant geographic expression in the distribution of land uses and infrastructure;

(b) the spatial logic of maquiladora (in-bond) industry;

(c) the characteristics of the tourist industry in Baja California;

2 The impact of transborder traffic congestion on the tourist industry and commercial activities which link the two cities;

3 The potential tourist demand for a new border crossing near the Pacific Ocean as a proportion of total transborder trips;

4 The level of efficiency of existing border-crossing infrastructure.

The conceptual framework of this study relies on the notion of an asymmetry of political and economic power relations between Mexico and the United States; this unequal relationship is reflected by the differing levels of urgency with which the two national subregions define the 'problem' of transboundary traffic congestion, which in turn produce divergent views on the location of a new border-crossing facility. Whereas in the United States the long lines at the crossing gates represent an irritant for North American tourists intent on visiting points of interest in Baja California, in Mexico the new border-crossing facility is perceived as a catalyst which will induce greater economic development for the region, an issue which acquires increasing relevance as both countries approach implementation of the Free Trade Agreement.

URBAN GROWTH IN A BINATIONAL CONTEXT: SAN DIEGO AND TIJUANA

Tijuana's urban development trajectory (1950–present)

Since 1950, Tijuana has emerged from its position as a small city, geographically and economically isolated from major national markets, to that of a large, highly urbanized entity of regional significance increasingly tied to the dynamic market of southern California.

In the immediate post-war period, Tijuana experienced some of the most dramatic population growth rates in the western hemisphere; whereas in 1950 it had a population of 65,364 by 1980 it had mushroomed to 709,340, representing more than a ten-fold increase (see Table 2.1).

Table 2.1 Tijuana population growth, actual and projected

Year	Population	Growth rate (%)	Percentage of population of Baja California
1950	65,364	12	29
1960	165,690	10	32
1970	340,583	7	39
1980	709,340	7	39
1985	867,719	6	47
1990	1,129,000	6	–
2000	1,815,000	5	–

Source: Gioberno del Estado de Baja California (SAHOPE) (1984), Plan de Desarrollo Urbano Centro de Población, Ciudad de Tijuana.

By 1985, nearly half the population of Baja California was concentrated in the greater Tijuana metropolitan area; migration flows originated primarily from the states of Jalisco, Michoacan, Sonora and Sinaloa (IIS–UABC 1984). Successive migration flows had a formative spatial impact on Tijuana's growth. Prior to 1950, the city's population was concentrated around an inner core near the traditional downtown commercial nucleus at the international border. As migrant waves intensified from Mexico's interior states, Tijuana's spatial form adapted in two directions:

1   migrants arriving with few economic resources were forced to take residence where land prices were relatively inexpensive, along the precarious ridges and ravines located in the southern and south-western perimeters of the urban core, and;
2   migrants with high levels of educational and professional achievement settled in exclusive residential zones near the principal arteries of the central city or in newly 'suburbanized' enclaves near the ocean.

The spatial growth of Tijuana has been principally shaped by the arrival of large numbers of low-income migrants who established their presence through land invasions or in areas where property rights could be obtained. Of the population growth occurring between 1950 and 1980, more than one half of the inhabitants occupied spontaneously settled 'colonias populares' (low-income neighbourhoods) (Herzog 1990: 182).

Tijuana's spatial transformation was also influenced by changes

20

taking place in its economic base resulting from its accelerating integration with the United States economy. Between 1950 and 1980, Tijuana's economy shifted its emphasis from primary and secondary activities to one in which the tertiary (service) sector became predominant (see Table 2.2). The most important of these emerging activities were retail and wholesale trade, tourism and services. Manufacturing activities also expanded and diversified, owing to the increased presence of the maquiladora industry, which drew US and Japanese investment capital to the region.

Table 2.2 Tijuana employment by sector, 1950–82 (percentage)

| | Percentage of employment | | | |
Sector	1950	1960	1970	1982
Primary	24.2	21.4	10.1	6.8
Secondary	23.8	27.2	32.3	33.9
Tertiary	52.0	51.4	57.6	59.3

Source: CONAPO (1984), Estudio Sociodemográfico del Estado de Baja California, Mexico.

Complementing efforts to revitalize and diversify the local economy, by the mid-1970s the Mexican Government attached increasing importance to the transformation of the Mesa de Otay, a flat expanse of land located east of the downtown centre near the international border. Two national decrees divided the land into a residential development zone and an industrial development area. The area surrounding the mesa was targeted by the federal government as a major residential–industrial 'growth pole' that would serve to extend the city's economic base away from the city centre, while providing future jobs for newly arriving immigrants (Flores 1990). At the heart of community plans for the mesa was the establishment of a large industrial park containing a major share of the city's maquiladora industrial activity.

San Diego's development trajectory: 1950–present

In the post-war period, San Diego emerged from its position as a small, peripheral city with an economy dependent on the presence of the US Navy to one of the twenty largest urban areas in the nation. San Diego's population increased from approximately 556,000 in 1950 to over two million by 1985, a growth of 274 per cent. This dynamic pattern of growth mirrored processes occurring in other cities of the US south-west subject to the demographic transition from Rustbelt to Sunbelt economies.

Much of the change in San Diego's spatial structure was conditioned by its expanding economic base. After the Second World War, the city's predominant employment sector continued to depend on the local navy establishment, the largest in the country, as well as a marine corps base at Camp Pendleton.

Initially, San Diego began to diversify in sectors of production closely linked to military activities, such as aircraft and aerospace manufacturing. It also invested in new areas of production, such as chemicals, apparel and electrical and non-electrical machinery. By the end of the 1970s, the city's manufacturing base comprised 15 per cent of its employment resources. During this period, tourism also became a dynamic growth industry, attracting over one billion dollars to the region annually and occupying 10 per cent of the area's employment base.

The locational preference of a large professional class of new migrants, coupled with patterns of local economic restructuring, had significant impacts on San Diego's spatial character. In 1950, the city's modest population was concentrated around the traditional urban centre located near San Diego Bay and adjacent neighbourhoods. Older communities such as La Jolla, Pacific Beach, Ocean Beach and the south bay cities of Chula Vista and National City were linked to the city core by trolley system.

Between 1950 and 1960, the direction of San Diego's growth was defined by the ascendancy of the automobile and the expansion of freeway networks across the urban landscape. These processes in turn encouraged urban deconcentration, which transformed the emerging spatial structure of the city into a multi-polar urban environment. In the decade spanning 1960–70, the areas of the city to the south-east continued to densify as processes of suburbanization continued. By the late-1980s, San Diego's spatial formation was spread over a broad patchwork of territory, interlinked by a sophisticated freeway network which promoted the formation of quasi-autonomous urban areas at the city's periphery. The areas of highest urban density include communities situated around the Interstate 5 freeway along the coast as well as the south bay corridor, which extends to the US–Mexico border.

TRANSFRONTIER INDUSTRIAL LINKAGES, THE PANACEA OF TOURISM AND BORDER-CROSSING FLOWS: TIJUANA AND SAN DIEGO

During the 1970s and early 1980s, the urban development paths of Tijuana and San Diego, though responding to different stimuli, converged spatially on the Otay Mesa/Mesa de Otay. Whether in the form of a 'New Town' community on the US side of the border or a 'growth pole' in Tijuana, the two sides of the mesa were considered as integral parts of each city's population growth-management efforts as well as platforms for future economic expansion and employment stimulation. The growing presence of US and Asian industries linked to Baja California through the maquiladora programme and the dramatic surge in tourist-related activities offer a conceptual framework with which to link the changing patterns of urban growth in both cities to the recent sharp increases in traffic volume at the existing border gates.

Industrial expansion on the Otay Mesa/Mesa de Otay

For the greater Tijuana/San Diego metropolitan area, the last decade has been characterized by an impressive expansion of interregional industrial linkages based primarily on the growth of the maquiladora programme. Originally established in Tijuana in the mid-1960s under the terms of a federally sponsored national border industrialization programme, maquiladora traditionally represented the labour-intensive, low-cost assembly component of a foreign firm's production ensemble transplanted to Mexico to reap the economic benefits of a less expensive and politically docile labour pool.

The 1980s witnessed dramatic changes in the nature of maquiladora production in Mexico; high rates of automation and product diversification now characterize many sectors of the industry in Baja California. Labour costs no longer dominate the economic agendas of many of the firms settling in Tijuana, as international competition within industrial sectors appears to be increasingly dictated by the requirements of production flexibility and quality rather than price. The geographic expression of this phenomenon is seen in the formation of broad transnational and transectoral alliances whose locational requirements are capital intensive. Asian firms in Tijuana over the last five years attest to the impact of global industrial trends on the San Diego/Tijuana region.

As of June 1988, 413 manufacturers were established under the maquiladora 'in-bond' programme in Tijuana, having doubled in number since 1986 (see Table 2.3)

The expansion of the maquiladora industry in Tijuana and the intensification of interregional linkages with the southern California economy have contributed to increased traffic flows at the border. Between 1985 and 1987, maquiladora plants increased their share of component parts and material imports by 70 per cent. In 1987, the value of imports passing through the San Ysidro border crossing exceeded $900 million. The third quarter of 1988 showed a 55 per cent increase over the same period in 1987.

A primary industry policy objective of Tijuana has been to channel maquiladora activity to the city's established industrial park in the Mesa de Otay. In the past few years, maquiladora activity in the mesa, particularly in the Ciudad Industrial Zone of 'New Tijuana', has mushroomed. In 1988 nearly one quarter of all maquiladora activity in Tijuana was occurring in the Mesa de Otay, consuming 42 per cent of all raw material imports entering through the San Ysidro port of entry and generating more than one third of maquiladora-related economic aggregate value added. Much of the dynamism of the maquiladora industry can be accounted for by the existence of a free trade zone encompassing the entire northern portion of the state of Baja California. Since its inception in 1973, the zone has created the conditions for future maquiladora expansion in Tijuana by waiving customs duties for the export of finished goods to the USA; once exported to the United States, duty is only levied on the value added by assembly or manufacture in Mexico. As a result, the free trade zone has provided an

Table 2.3 Tijuana maquiladora industry, 1986–8

Date of register	Number of firms	Total percentage
Before 1986	212	51
January to June 1986	31	
July to December 1986	40	
January to June 1987	35	49
July to December 1987	45	
January to June 1988	50	
Total	413	100

Source: Subdelegación de Fomento Industrial (SECOFI), Tijuana, B.C.: Mexico, October 1988.

important economic and social link between Tijuana and southern California.

Within San Diego city limits, in the Otay Mesa, much of the development activity in the past few years has been characterized by US and Japanese industrial sites and warehousing facilities established to service affiliate plants across the border in the Mesa de Otay in Tijuana. An examination of the nature and intensity of physical development over time in the Otay Mesa confirms an awkward and fitful period of growth between 1987 and 1989, followed by a dramatic increase in industrial and commercial activity up to the present time. Whereas in 1987, 916,462 square feet were developed in the Otay Mesa, only 506,014 square feet of land were absorbed in 1988. An additional 525,545 square feet were amassed in 1989, representing a modest growth of 3.8 per cent. Yet, for the projects planned for completion in 1990, 630,214 square feet were allocated, representing nearly a 20 per cent increase from the previous year.

Although the proportion of industrial/warehouse activity to total land uses had been declining from a high of 80.1 per cent to 70.9 per cent in 1989, in 1990 nearly 96 per cent of land acquired for development was dedicated to such uses, a dramatic indication of the extent to which industrial activity in Mexico is being mirrored on the US side of the border.

Of a total of eight-six firms operating in the Otay Mesa in 1989, nearly one third relied on affiliate plants in Tijuana for the final elaboration of their products. The rate of growth of Otay Mesa firms with maquiladora links in Tijuana is impressive; between 1987 and 1988 employment in such companies increased by 771 per cent.

CONGESTION AT THE BORDER GATES: IMPLICATIONS FOR THE VOLUME OF TRANSBORDER FLOWS

Tourist traffic flows

A large component of traffic flows at the border is attributable to tourism. The tourist industry, one of the principal generators of revenue for the city of Tijuana, is estimated to have generated over 1 billion dollars in 1988; between 1984 and 1988 tourists' expenditures in Tijuana increased by 215 per cent (Bringas 1988: 7). Tijuana remains the main point of attraction for foreign tourists to Baja California, capturing three-quarters of all border crossings in the last five years

(Bringas 1983: 33). A 1987 survey by the Colegio de la Frontera Norte determined that 57 per cent of tourists visiting Tijuana are of Mexican origin, 29 per cent are Anglo-American (White) and 14 per cent Asian, European and Black.

Border-crossing activity

The San Ysidro border crossing, the main port of entry for southbound tourists, is a twenty-four lane facility which processes an average of 63,000 vehicles per day. In recent years, the San Ysidro border crossing has experienced the highest volume of population and vehicle flows in its history, registering a 21 per cent increase between 1987 and 1988 and a 17 per cent rise between 1988 and 1989 (US Immigration and Naturalization Service 1989).

Since the mid-1980s, the proportion of Mexican crossers has remained stable at 54 per cent, comprising a majority among the users of the San Ysidro facility. The proportion of non-US citizens crossing the border peaked in the late-1970s and early-1980s, dropping by 6 per cent in the year following the 1982 peso devaluation. As a consequence of the Immigration Reform and Control Act of 1986, it is expected that increasing numbers of newly legalized Americans of Mexican origin will return to Tijuana to visit family and friends.

Border-crossing activity at the Otay Mesa port of entry has grown parallel to that at San Ysidro. Between 1987 and 1988, crossings at the mesa jumped from approximately 6 million to 7.2 million, representing a 20 per cent increase over the previous year. The Otay Mesa border crossing, initially established to decongest traffic conditions at the San Ysidro facility, has not achieved its goal; between 1987 and 1988 traffic volume at San Ysidro increased over five times as much as at Otay Mesa.

Impact of transborder congestion on industrial activity

Of the firms located on the US side of Otay Mesa with linkages to the maquiladora industry in Tijuana, 71 per cent considered the border-crossing delays at the Otay Mesa port of entry a 'problem'. The average waiting time recorded by firms expecting shipments through the Otay Mesa crossing was five hours. In every instance where the problem of border traffic was mentioned, the principal concern was the inordinately long inspection procedures at the Otay Mesa port of entry.

Delays in commercial traffic processing time were attributed mainly to arbitrary inspection procedures at the border-crossing gates and an

26

inadequate number of customs personnel to handle commercial border flows. The survey results demonstrated that traffic delays at the Otay Mesa port of entry have a largely negative impact on the capacity of firms to conduct business on both sides of the border; the 'friction of distance' occasioned by the political border between Mexico and the United States, while preserving the binational division of labour and capital between the two countries, is nevertheless not porous enough to satisfy industrial transaction requirements fuelled by the accelerating economic integration of the two national sub-regions.

Impact on Tourism

Representatives of the tourist industry on both sides of the border have demonstrated frustration at the border-traffic delays at both border crossings. In San Diego, the president of the Convention and Visitor's Bureau stated he was 'surprised' to discover a 20 per cent increase in border crossings over the previous year. Employees of the Bureau further indicated that, because of the dependence of US commercial centres on Mexican consumers, the border-crossing delays have a retail impact on the local economy.

In 1988–9, tourist industry representatives along the coast of Baja California formed a lobbying group to promote the establishment of a third border crossing between Tijuana and San Diego. Under the leadership of the Tijuana Tourism and Convention Bureau, the Committee to Expedite the Border Crossing, otherwise known as COPAC, actively sought to promote a border crossing near the Pacific Ocean. A COPAC member representing the Mexican National Chamber of Commerce of Restaurants and Hotels (CANIRAC) explained: 'This (the traffic delay at the border) is not a problem of Mexico, yet in the minds of most tourists it is associated with the Mexican portion of their trip, and leaves a bad aftertaste in their memory.'[1] Broadening the regional implication of the border-crossing delays within the framework of the binational trade relationship, one CANIRAC official stated:

> Under the terms of the commercial opening implied by GATT, which Mexico is soundly adhering to, it is incomprehensible that the U.S. border crossing is not adapted to the commercial requirements of the San Diego–Tijuana border zone. The North American government has been negligent in this respect in complying with its part of the GATT agreement.

27

It is doubtful that the speaker meant to connect GATT to the border-crossing issue in a direct manner, as the two issues are not linked in any official policy statement. The reference does indicate, however, the degree of frustration in Mexico at the inadequacy of border-crossing facilities in the border area.

As a response to the demands from the tourist sector in Mexico to open a third border crossing by the sea, it is necessary to determine the mobility patterns of all the potential users crossing the border in order to anticipate the best location and the scale of infrastructure needed for a new port of entry. Assessing the quantitative nature of development pressures impinging on both sides of the border may lead to the act of building a new border crossing, but the location of such a facility should be defined by the needs of the population demonstrating the greatest potential demand.

Analysing the spatial activity of tourists by ethnicity in the context of total border-crossing flows allows the gauging of the level of demand for a border crossing by the sea. Whereas 77 per cent of the Anglo-American tourist population considered Tijuana as a final point of destination, 93 per cent of the Mexican-American visitors came to spend their time exclusively in the border city (Bringas 1988: 32).

In applying ethnic percentages to the total annual volume of US citizen transborder flows, it becomes evident that the Mexican-American community is the primary tourist constituency visiting Baja California. In 1988, for instance, over 12 million Mexican-Americans entered Baja California via Tijuana, compared to only 6.1 million Anglo-Americans and 3 million Asians and Blacks. As a result of the legalization provisions of the Immigration Reform and Control Act of 1986, it is expected that the Mexican-American tourist constituency will absorb an even larger share of annual San Ysidro border crossings. Anglo-American tourists comprise the largest group of visitors to points south of Tijuana. In 1988, for instance, over two million Anglo-Americans ventured further south than Tijuana, whereas only 853,435 Mexican-Americans did likewise in the same period. Yet, compared to the aggregate of tourists visiting Baja California, the number of Anglo-American visitors travelling further south comprises only 9.5 per cent, as compared to 53 per cent among the Mexican-American tourists who remain in Tijuana proper. The number of southbound Anglo-American tourists is further reduced as a proportion of total border crossers (US citizen and non-US citizen), representing only 4 per cent of total volumes passing through the San Ysidro facility. Conversely, Mexican-American visitors constitute nearly 27 per cent of total border-crossing flows.

28

The primary constituency using the San Ysidro border crossing on a day-to-day basis remains a large local commuter population composed of Mexican labourers bound for San Diego and surroundings cities. For this group, San Ysidro will continue to be the most accessible port of entry into the United States because of its proximity to Tijuana's urban core and ancillary transportation routes. Of the tourist population visiting Tijuana, the Mexican-American population is undeniably the principle user; in this context it is doubtful that a crossing on the coast would improve their access to the centre of Tijuana or the hillside 'colonias' on the urban periphery.

ANALYSIS OF BORDER-CROSSING LEVELS OF EFFICIENCY

The pace of border-traffic flows at San Ysidro and Otay Mesa is determined largely by the number of gates operational at any particular time. Given the high levels of traffic congestion at both ports of entry, the levels of efficiency may serve as useful criteria in assessing the opportunity costs of building a new border-crossing facility; to the extent that existing border crossings are under-utilized, alternatives to building a new crossing present themselves.

The San Ysidro port of entry, a structure with a total of twenty-four gates, maintains three gates closed at all times: two on the outer perimeters of a wide entry funnel and one along the centre of a median strip. In theory, twenty-one lanes should be available to traffic at full capacity. The crossing at Otay Mesa has the ability to use its full ten-lane infrastructure if the need arises. A study of lane utilization at both border-crossing gates revealed a strong inverse correlation between the supply of gates and the number of cars per line waiting to pass into the USA.[2] During the week in which measurements were taken, at the San Ysidro border crossing an average of only fourteen gates were operational at any one time, representing only a 66 per cent level of capacity. Even during peak flow times, a maximum of sixteen gates were functioning, representing 76 per cent capacity. Inefficient use of the Otay Mesa border-crossing facility was also evident; with a potential supply of ten gates, a weekly average of six were operational, representing a 60 per cent capacity ratio.

SUMMARY OF FINDINGS

San Diego and Tijuana have responded to their internal demographic

29

and industrial growth pressures by reshaping the structure of permissible land uses to accommodate and channel development in the direction of their respective peripheries adjacent to the international border. Processes of binational economic integration, which may be confirmed by Mexico and the United States through the signing of a Free Trade Agreement, will accentuate regional economic inter-dependencies within the San Diego–Tijuana urban area. The visible increases in traffic congestion at the Otay Mesa and San Ysidro border crossings are a product of forces that can only be apprehended regionally, involving

1 the expansion of industrial ties between San Diego- and Los Angeles-based firms linked to the maquiladora industry in Tijuana,
2 a burgeoning growth in the American tourist industry centred on the Tijuana border community, and
3 the heightened flow of Mexican commuters crossing into the USA for shopping and employment opportunities.

Yet, despite the consequent strains of border-traffic congestion which emerge as a result of increased transboundary regional economic linkages, the two existing border-crossing stations continue to operate below capacity, demonstrating an apparent lack of coordination between the two governments in solving a transportation problem whose negative externalities are perceived on both sides of the international dividing line.

Several recommendations flow from the preceding analysis. The construction of a new border-crossing facility between Tijuana and San Diego near the Pacific Ocean would not be justified based on the following observations:

1 The highest rates of population and industrial expansion in Tijuana are occurring in the eastern and south-eastern portions of the city and, as a result, the pressures for increased transborder access are perceived in the eastern sectors of the city, not near the ocean;
2 There would not be a sufficient demand to justify the expense incurred in the construction of a new border crossing; as a proportion of total transborder crossers annually, the tourist market attracted to a coastal port of entry would represent a minimal percentage of total border traffic flows;
3 Owing to the under-utilization of the San Ysidro and Otay Mesa gates, the number of border-crossing officials at the two ports of entry should be increased in order to maximize the use of existing facilities.

NOTES

1 Personal communication of CANIRAC representative, January 19, 1990, Tijuana, B.C.
2 The study of border-crossing lane utilization was conducted at the San Ysidro and Otay Mesa ports of entry between April 2–6, 1990.

BIBLIOGRAPHY

Bringas Rabago, N. (1988) *Turismo Fronterizo: Caracteristicas de los Visitantes de Origen Mexicano y Angloamericano en Tijuana,* El Colegio de la Frontera Norte, Departmento de Estudios de Administración Publica, Tijuana, B.C.

CONAPO (1984) Estudio Sociodemografico de Baja California, Mexico.

Flores, A.C.G. (1990) (Director of Land Use Planning) SAHOPE–Mexicali, Personal communication, January 3.

Gobierno del Estado de Baja California (SAHOPE) 1984, Plan de Deserrollo Urbano Centro de Publación, Ciudad de Tijuana.

Herzog, L. (1990) *Where North Meets South: Cities, Space, and Politics on the United States–Mexico Border,* Austin: University of Texas Press.

U.S. Immigration and Naturalization Service (1989) San Ysidro border crossing statistics provided to author: 1954–1989, San Ysidro, CA, October.

3

MYTHS AND REALITIES OF TRANSBORDER POLLUTION BETWEEN CALIFORNIA AND BAJA CALIFORNIA

Cuauhtémoc Leon and Marina Robles

INTRODUCTION

The study area is located in one of the most arid regions of the planet, in the north-west portion of Mexico and the south-west of the United States. This area comprises the biogeographical region of the Californias, which encompasses a portion of the richest state in the American Union, California, and perhaps the most conflictive peninsula in Mexico, Baja California, where this study is focused. Up to 1848 this region belonged to a single entity, not only from a biological perspective but also from a geopolitical point of view, since California formed part of Mexico. Since then, Baja California has remained a part of the Mexican Republic, and California has been a part of the United States of America.

A biological region usually offers characteristics of relative geographic and climatic homogeneity which enable species of fauna and flora to disperse and exchange their respective gene pools unhampered by boundaries (Brown 1982). Greater California is a desert region, with a central mountainous system which separates the Mediterranean climatic belt along the coast from the subtropical desert climate inland. As a rule, summers are dry and extremely hot (up to 50°C) and winters tend to be cold (down to 0°C) and rainy for the coastal regions (Wiggins 1980). Mean precipitation levels average 250 mm per annum (Roberts 1989), and interannual variations can be significant. Periods of droughts lasting up to sixty years have been recorded in the region (Lenz and Dourley 1981). This fact reinforces the importance of adequate water-resource management policies in such arid regions.

WATER RESOURCES AND POPULATION

Freshwater sources in the region are essentially confined to two: the Colorado River Basin and the Northern California Aqueduct, which supplies most of the southern portion of the State of California. The Colorado Basin waters are utilized on both sides of the boundary, although the largest proportion is consumed in the USA, while the delta flows into the Sea of Cortés on the Mexican side of the border.

The presence of this transboundary watershed has been at the root of competition and conflict over its use since the mid-nineteenth century. Tensions over water-resource use were partly resolved by the signing of the 1944 Treaty on Boundaries and Water Resources between the two countries. This treaty provided for a fixed quota of 1,850 million cubic metres of water per annum for Mexico. On the other hand, California was awarded a far larger quota of 5,426 million cubic metres of water per annum, practically three times as much (Graft 1985) (see Table 3.1).

While constituting an important source of freshwater for California, the Colorado River is not its principal supplier. California's main sources of water are the Aqueduct in the north, which provides seven times more water than the Colorado River, and groundwater extraction which supplies it with four times more water than the river it shares with Mexico (United States Geological Survey 1985, cited in United States Department of Commerce 1990).

California is the most populated state in the USA with close to 30 million inhabitants, and is at the forefront of the country's industrial and agricultural production (op. cit.). This was made possible largely thanks to state and federal hydraulic works which enabled the transport and storage of water from different locations. The availability of large volumes of water where it was previously absent explains to a large degree the massive environmental transformations undergone in many areas of southern California.

The most important local source of freshwater in Baja California is the Colorado River, the quota for which was fixed as indicated above by the 1944 treaty. This quota, added to the volume of groundwater extracted from the Colorado Delta region, contributes up to 80 per cent of available water resources for the state of Baja California. These sources still do not supply the population with enough to cover its water needs (Rojas-Caldelas 1991).

As a corollary to the Colorado River's transborder influence, marine currents connect ecosystems from north to south. The California Current, off the Pacific Coast of the region, drives cold waters from the

Table 3.1 General statistics for California and Baja California, 1990

	Baja California	California
Total population	1,657,927 (2% of national population)	29,063,000 (11.2% of national population)
Volume of waters from the Colorado River	1,850 m. m³	5,426 m. m³
Water consumption	3,250 m. m³	21,100 m. m³
Surface waters	1,850 m. m³	34,600 m. m³
Groundwaters	1,400 m. m³	15,100 m. m³
Native species (including varieties)	4,000+	7,700
Endemic species of flora	700*	3,673
Endangered species of flora	23*	950
Endangered ecosystems	Coastal lagoons Canyons Dunes Grasslands	Coastal lagoons Canyons Dunes Grasslands
Environmentalist agencies and groups	13	145

Note: *estimated figures.
Sources: various.

north along the Baja California coastline. The importance of this current lies, on the one hand, in its high productivity in pelagic resources but, on the other hand, in its capacity to transport pollutants. The dominant winds in the region are north-westerlies, which also carry airborne pollutants from the north (see Figure 3.1).

WATER QUALITY

California has implemented widespread urban reforestation programmes for some years, in contrast to the total absence of similar measures on the Mexican side of the border. However, the extensive planting of trees has caused serious setbacks because of high water consumption, which competes with other priority needs. In most cases, the species of trees planted come from more humid regions and have high requirements of water.

However, by far the largest consumer of water in California has been the agricultural sector. The development of irrigated agriculture has

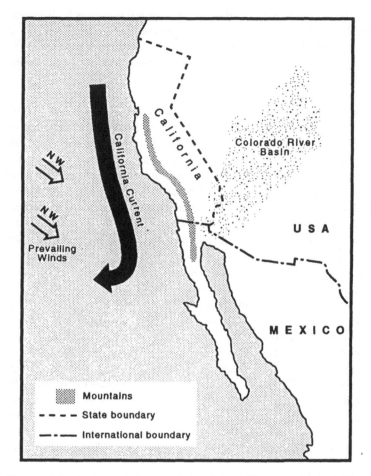

Figure 3.1 Environmental characteristics of the California and Baja California region

depended essentially on groundwater extraction from transboundary aquifers such as those located in the Colorado basin, which has had direct impact on the availability of water on either side of the border. Furthermore, the high concentrations of salts and pesticide residues in the waters of the Colorado River have not only affected water quality but have also caused serious soil deterioration on the Mexican side (Alvarez-López 1988).

Similarly, seawater quality is directly affected owing to the north-south trend of the California Current. Since most of the freshwater used

35

for California's urban, industrial and agricultural purposes finds its way to the Pacific Ocean, many coastal settlements on the Mexican side of the border which depend on fishing activities are directly affected. In an attempt to analyse the interrelation between water quantity and quality, a simplified model of water use in California and Baja California is presented in Figure 3.2.

The border region contains the largest population centres of both states. Concentrated between the cities of Los Angeles and San Diego is over 40 per cent of the population of the State of California (approximately twelve million inhabitants), and over 80 per cent of the population of Baja California can be found between the cities of Ensenada, Tijuana and Mexicali. California's urban centres dispose of their domestic and industrial waste waters through submarine offshore conduits. In Baja California, most urban and industrial waste waters are disposed of directly, and in the case of Tijuana and Mexicali these are evacuated into northbound rivers which cross into California (Alvarez-López 1988) (see Figures 3.3 and 3.4).

Figure 3.4 illustrates the relative importance of waste waters containing heavy metals in the oceanbound discharges of each state. Waters containing traces of heavy metals are considered highly toxic, and in high concentration they cause irreversible biological impacts on the oceans' food chains. Comparing the combined toxic discharges of Tijuana, B.C. and San Diego, CA, it is apparent that San Diego contributes up to 75 per cent of the total toxic substances, including 95 per cent of cadmium and 85 per cent of mercury emissions (Segovia-Zavala, et al. 1986), elements considered lethal for living organisms.

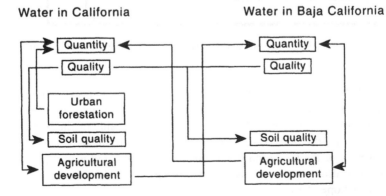

Water in California   Water in Baja California

Figure 3.2 Influence of water quantity and quality on different activities and soil quality

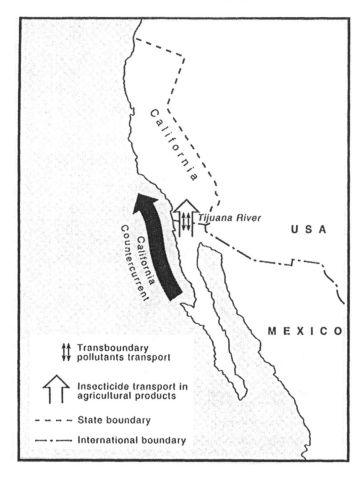

Figure 3.3 Natural and man-made countercurrents

On the other hand, Tijuana is responsible for over 50 per cent of lead emissions (see Figure 3.5). This provides us with an idea of the magnitude of California's contribution to coastal pollution, which reflects the high degree of industrial development along the southern California coast. Most of the toxic elements discharged offshore are eventually transported southwards by the California Current and accumulated in the tissues of marine organisms which are in turn extracted by the fishing industries on either side of the border.

Some gross calculations suggest that California north and south of the border introduces between eleven and seventeen tonnes of mercurial

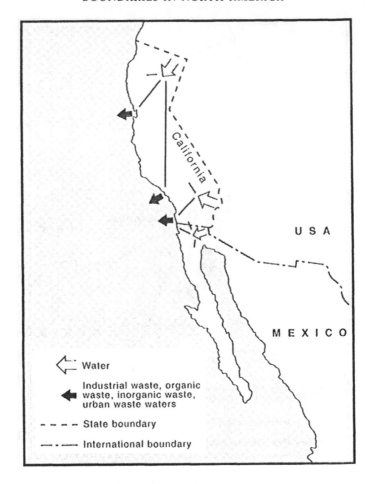

Figure 3.4 Water sources and distribution and foci of waste products in Greater California

compounds per year into the Pacific Ocean Basin (spanning from Punta Concepción, USA, to Punta Colonet, Mexico). At least 74 per cent of these pollutants are fallout from airborne emissions, and 25 per cent come in the form of waste water discharges from the principal metropolitan centres in southern California; the rest is contained in surface run-off waters (Gutierrez and Flores 1986).

38

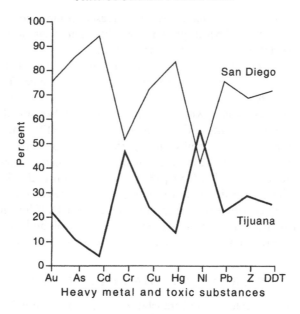

Figure 3.5 Percentage participation of San Diego, California and Tijuana, Baja California in the discharge of toxic substances
Source: adapted from data in Segovia-Zavala *et al.* 1986

TRANSBOUNDARY EXCHANGE ZONES

In addition to seasonal variations in atmospheric and ocean currents that produce small-scale transport from either side of the boundary, there are two other northbound currents, one natural, the other artificial. The first is the California Countercurrent which occurs at depths below 200 metres and runs in the opposite direction to the California Current. Due to its depth, most superficial urban and agricultural discharges from Mexico do not reach California, except during short periods of the year when these conditions vary. The other, man-made, contamination vector is found in the transport of insecticides and other toxic substances that cross the border into the United States from Mexico, in the form of agricultural and food products. These contaminated products are often screened by the US Food and Drug Administration and returned to their origin (see Figure 3.3).

During the 1980s, seafood consumption in California increased considerably, and a large proportion of Baja California's exports of seafood is bound for its northern neighbour. The growing trend of vegetarianism in California leads to increased competition between locally-grown and Mexican-grown vegetables and fruit, a source of concern in California's agricultural sector (Fay 1987).

California's population is seventeen times greater than that of Baja California, and consumes twenty-one times more water. This fact, combined with a high technological capacity for environmental engineering, may explain why there is a greater number of endangered species and environments in California than in Baja California.

Moreover, California ranks among the three US States with the largest number of toxic waste disposal sites, with ninety-one such sites scattered across the region (Fay 1987). The environmental liabilities of these sites have been the subject of controversies between the state and university and environmentalist groups (see Table 3.1). While increased environmentalist militancy has led to increased controls and monitoring, the overall results in terms of curbing pollution are still questionable.

A FRAGILE TRANSBOUNDARY ECOLOGICAL WEB

Faced with a complex web of transboundary pollution, one is tempted to believe that each source of contamination is separate and independent. However, in practice, nature ignores political boundaries, and the reproductive strategies of many species maintain unalterable links between California and Baja California. An intricate system of ecological interaction exists between the Californias, due to the absence of biogeographical boundaries, and allows the interaction of distinct populations of species at varying intensities. Distinct species adopt particular reproductive strategies. For instance, certain species of marine and terrestrial flora adopt similar strategies to transport spores and seeds, using sea or wind currents for dispersal. Political boundaries are non-existent when it comes to the dispersal, colonization and reproduction of these species. Thus one can conceive the border region between California and Baja California as an immense web of ecological interaction.

Fire controls provide a concrete example and a good illustration of the differentiated environmental impacts on either side of the border. Fire prevention programmes in California have produced a different vegetation community. Over a century of fire and disaster controls have modified the spatial dynamics of the reproduction of species of flora, thus transforming the natural landscape. South of the border, where forest fires are not extinguished, affected areas are small and produce a mosaic of young and adult plants, contiguous to fire-prone areas of old or dead vegetation. Fire is usually detained by patches of young and green vegetation which have not accumulated sufficient inflammable material. The mean area destroyed annually by fires in Baja California averages 1,600 hectares, less than half the figure for its northern neighbour in the USA (over 4,000 hectares). The negative effects of the fire prevention programme are associated with the natural cycle of the vegetation community which often requires fires for the regeneration and succession of species. This often means that, due to lack of fires, several fire resistant species cannot survive, and are overwhelmed by dominant species (Delgadillo 1992). Several studies point to the need to develop strategies and actions which imitate fire events in order to enhance species succession. Finally, findings suggest that, while fire prevention policies in California tends to reduce the frequency of these events, the magnitude of these fires tends to increase, leading to devastating and dangerous events for urban areas, as has been the case recently (Minnich 1983).

Biological divergences

Urban sprawl, agricultural activities and hydraulic works have a direct impact on the immediate environment in which they occur. Indirect impacts frequently affect much wider areas, and their intensity does not necessarily diminish with distance. Deforestation and groundwater extraction lead to habitat changes. With the disappearance of a vegetation community or habitat, as in the case of coastal lagoons or canyons, many species of insects, reptiles, birds and mammals associated with it either migrate or vanish. If the habitat constituted a reproduction site, or a stopover site for migratory birds, or if it contained particularly vulnerable vegetation species, its modification would affect a wide range of lifeforms simultaneously. Moreover, evidence suggests that the introduction of exotic vegetation species in forestry projects tends to displace local species (Graft 1985).

Finally, several case studies are analysed in order to illustrate the divergent tendencies present in the border region and the mutual influences affecting either side (see Figure 3.6).

Case No. 1 A species disappears on either side of the border

1   The Fremontia Shrub (*Fremontodendron Mexicanum*), which was originally restricted to the area of San Diego–Tijuana and associated with the Chaparral ecosystem, is now considered extinct

41

due to urban sprawl and deforestation (Lenz and Dourley 1981).

2   By the end of 1800, the last remaining bear (*Ursus horribilis*) still in the region was hunted. Today, this species is found only in some portions of the northern United States and in isolated areas of Mexico (Starker 1985).

Case No. 2 A species (or environment) exploited in one state without having an impact on the other

2.1 Exploited or disturbed environments in the USA

1   In California, the high rate of exploitation of abalone species (*Haliotis* sp.) which live on rocky shores and of a Clam species (*Tivela stultorum*) which thrives in sandy environments had all but depleted the state of these species by 1974 (Bonnot 1948 and Morris *et al.* 1980). In Baja California, these species are still undergoing intense extraction, and their permanence is solely due to the fact that their exploitation is more recent. In California, the exploitation of these species was regulated, and they seem to be recovering progressively.

2   Fire controls in California's chaparral and forest have modified the natural composition of the system and its dynamics. Fire naturally regulates the size and age of forest patches. In Baja California, this system is much less altered.

3   In Baja California, there are three important coastal lagoons, which, though modified by tourism and agricultural development, have functioned as a refuge for several bird species whose populations to the north have been strongly affected by the alteration or destruction of their reproductive habitats. The Least Tern (*Sterna albifrons brownii*) (Palacios 1992), which is listed as an endangered species in California, nests in these lagoons and has a stable population.

2.2 Exploited and disturbed environments in Mexico

1   In Baja California, traps placed along the banks of the Colorado River to catch beavers (*Castor Canadensis*) for their fur, led to their disappearance early in the present century.

2   On the Pacific coast, Russian and North American fur hunters exterminated the local populations of sea otter (*Enhydra lutris*) by the end of the last century.

3   Both the beaver and the sea otter populations of Baja California

suffered a similar fate, and are now extinct. In California, however, these species were reintroduced and protected through costly repopulation programmes.

Modelling environmental interaction on the border

Modelling interactions in a qualitative manner allows one to visualize the complexity of ecological linkages, identify the source and the receptor; that is, the direction of vectors. Modelling these interactions quantitatively allows one to measure the intensity of each vector, the vulnerable elements and the key or regulatory relations within the ecosystem, and therein lies the importance of this exercise; to identify where to intervene and where not to.

Clearly, it is difficult to learn from one's neighbour's mistakes, and even more difficult when basic needs are urgent and leave no time to wait for and seek a better solution. While now being 'trendy' and attractive, ecological variables are still but indicators of a complex whole. Their influence on the decision-making process still remains superficial, and they are still prone to erroneous use and interpretations. Nonetheless, there is a growing number of policies and projects which demand the incorporation of the environmental dimension in planning and policy making.

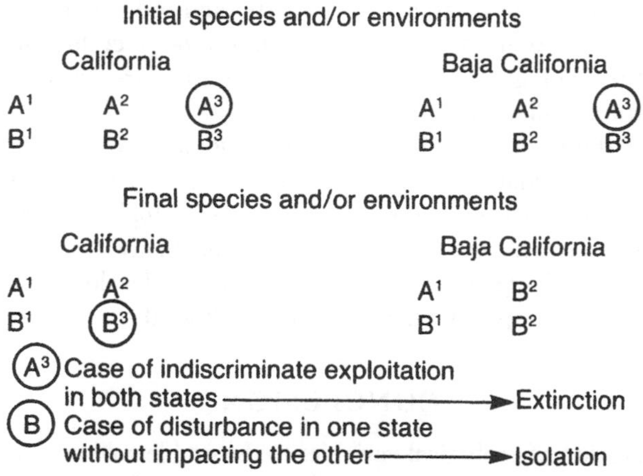

Initial species and/or environments

California   Baja California

A¹   A²   Ⓐ³   A¹   A²   Ⓐ³
B¹   B²   B³   B¹   B²   B³

Final species and/or environments

California   Baja California

A¹   A²   A¹   B²
B¹   Ⓑ³   B¹   B²

Ⓐ³ Case of indiscriminate exploitation
   in both states ──────────────► Extinction
Ⓑ Case of disturbance in one state
   without impacting the other ──────► Isolation

Figure 3.6 Simplified model of biological divergence possibilities between California and Baja California original environments and/or species

The economic model adopted by so-called under-developed countries supposes that, by following a predetermined productive strategy, they will attain the superior stages of growth in linear evolution, of which developed countries are an example. However, several writers have denounced this as a fraud and assert that the concept of development must be analysed from several critical points of view (Esteva 1985).

Production for production's sake, without a reflection as to why, has resulted in serious problems throughout the 1960s and 1970s. For instance, the 'green revolution' in Mexico caused several native strains of maize to disappear in favour of imported hybrids. A similar process occurred in India for rice. The massive conversion of lowland tropical forests into pasture for beef to export has impoverished more Latin American countries than any other activity. And there are endless numbers of examples of how large-scale development processes have resulted in environmental degradation, misery and a deepening of the state of under-development in which most of humanity lives.

Baja California, like its upper Californian neighbour, concentrates most of its urban structures close to the coast and the international border zone. This is the area with greatest potential for tourism, but it is also the area with the least available water. Urban sprawl is synonymous with growing water needs. The limited availability of the precious resource in Baja California means that it will impose insoluble restrictions on future growth. The choice will be between supplying water for tourism projects or for domestic and industrial use (Rojas-Caldelas 1989). Alternatives such as desalinization, recycling, improved irrigation technology and environmental education are urgently needed to face future challenges.

One of the traditional postulates of economic development is that there are unavoidable social and environmental costs. For Latin America, these costs have often been extremely high, and they have often compromised the future potential of many countries. The environmental degradation witnessed in many developed countries constitutes an example of what is not to be followed.

CONCLUSION

The application of ecological variables to the problem of transboundary interaction opens a new and challenging area of research. In this chapter, we have tried to contrast the environmental situation between California, USA and Baja California, Mexico. This border region

44

combines a highly contrasted economic and social environment with a relative homogenous and interactive biological environment.

However, examples of distinct fire control policies and differentiated environmental alteration through forestry projects and agricultural uses contribute to the growing biological divergence between California and Baja California. The zones and vectors of interactions between biological communities on either side of the border contribute to a complex web of ecological linkages. While producing divergent results, they tend towards unity.

The economic development models of each country have produced a differentiated use of resources. This in turn has favoured divergent distribution patterns by common species to distinct environments. The study of environmental dynamics across a political boundary from a historical perspective provides a key insight into the potential for common-policy making and shared responsibilities between neighbouring countries.

BIBLIOGRAPHY

Alvarez-López, J. (1988) 'El medio ambiente en el desarrollo económico de la frontera norte de México', *Cuadernos de Economía*, 5, 5: 1–106.

Bonnot, P. (1948) 'The abalones of California', *Fish and Game*, 34, 4: 141–69.

Brown, D.E. (ed.) (1982) 'Biotic communities of the American Southwest, United States and Mexico', *Desert Plants*, special issue 4, 1–4: 1–342.

Delgadillo, J. (1992) *Florística y Ecología del norte de Baja California, Serie Textos Méxicali*, Baja California, Mexico: Universidad Autónoma de Baja California: 339.

Esteva, G. (1985) 'Detener la ayuda y el desarrollo: una respuesta al hambre. Ponencia presentada en el Seminario Internacional sobre autosuficiencia alimentaria', unpublished document, Mexico: CEESTEM–UNESCO.

Fay, J.S. (1987) *California Almanac*, Los Angeles, California: Pacific Data Resources, 3rd edition.

Graft, W.L. (1985) *The Colorado River. Instability and Basin Management*, Research Publication in Geography, Arizona State University.

Gutierrez, E.A. and Flores, G. (1986) 'Disponibilidad biológica de mercurio en las aguas de la costa norte de Baja California', *Ciencias Marinas*, 12, 2: 85–98.

INEGI (Instituto Nacional de Estadistica, Geografia e Informatica) (1990) 'Resultados preliminares', *XI censo general de población y vivienda*, Mexico City: INEGI.

Lenz, L. and Dourley, J. (1981) *Californian Native Trees and Shrubs for Garden and Environmental Use in Southern California and Adjacent Areas*, California: Rancho Santana Botanic Garden.

Minnich, R.A. (1983) 'Fire mosaics in southern California and northern Baja California', *Science*, 219: 1287–97.

BOUNDARIES IN NORTH AMERICA

Morris, R.H., Abbot, D.P. and Haderlie, E.C. (1980) *Intertidal Invertebrates of California*, San Francisco: Stanford University Press.
Palacios, E. (1992) 'Anidación del gallito marino californiano (Sterna Albifrons) en baja California: Su relación con gradientes ambientales y de disturbio, e implicaciones para el manejo', MS thesis, Department of Ecology, CICESE, Universidad Autónoma de Baja California.
Paredes, A.E. (1985) 'Agua, recurso natural más importante para el estado de Baja California, México', Mexicali: Abril.
Roberts, N.C. (1989) *Baja California Plant Field Guide*, Los Angeles: Natural History Publishing Company.
Rojas-Caldelas, R.I., (1991) ' El Río Colorado, y el valle de Mexicali', *Ciudades*, 3(10): 33–8.
Segovia-Zavala, J.A. and Delgadillo-Hinajosa, F. (1986) 'Diagnóstico y alternativas de reducción y control de la disposición de aguas residuales sobre la zona costera fronteriza (100 km) México', in J. Alvarez y V. Castillo (eds) *Ecología y Frontera*, Coord. Tijuana, Mexico: Universidad Autónoma de Baja California: 236–44.
Smith, J.P. (1987) 'California's endangered plants and the CNP's rare plant program', in Elias, T.S. (ed.) *Conservation and Management of Rare and Endangered Plants*, Sacramento, California: California Native Plants Society, 1–6.
Starker, L. (1985) *Fauna Silvestre de México*, Mexico City: Editorial PAX.
United States Department of Commerce, Bureau of the Census (1990) *Statistical Abstract of the United States: The National Data Book*, Washington, DC: Government Publications Office.
Wiggins, I.L. (1980) *Flora of Baja California*, Stanford: Stanford University Press.

Part II
BOUNDARIES IN CENTRAL AMERICA

Part II

BOUNDARIES IN CENTRAL AMERICA

4

BORDER REGIONS IN CENTRAL AMERICA

An agenda for future research priorities

Allan Lavell

INTRODUCTION

The Central American isthmus (including Belize and Panama) is crossed by ten national borders with a total length of approximately 3,250 km, 25 per cent longer than the border between the United States and Mexico, and more than twice the length of the boundary between Colombia and Venezuela. At the same time, these borders cut across regions of intense economic and social activity, which in many cases have been subject to prolonged geopolitical disputes and in certain cases the setting for recurrent armed conflicts.

In spite of the social importance of border regions and of the changes that have occurred in these due to the economic and political crises Central America has suffered over the last decades, little or no systematic research has been conducted in the region concerning this subject. Nor do institutions exist in the region dedicated to the promotion of research efforts on this subject matter. Applied research promoted during the 1960s and early-1970s by a number of international organizations (IDB (Interamerican Development Bank), SIECA (Secretary for Central American Economic Integration), etc.) related to the quest for economic integration in the region provided valuable knowledge in terms of the potential for physical integration and for the timely use of shared natural resources. The bi- and multinational schemes which emerged at the time have not been updated or substituted since. The conflicts of the late-1970s and the 'lost decade' of the 1980s resulted in severe setbacks for the historical project of Central American integration.

Given the diversity and complexity of border regions seen from a Central American perspective, the setting of research priorities of a

49

regional comparative type requires a stage of theoretical and conceptual reflection and a systematization of existing studies and information on the subject. In the first instance, the definition of guidelines, theses and hypotheses allowed the organization and to a certain extent the homogenization of analytical frameworks in a systematic effort. These guidelines favoured the establishment of common research priorities for each border context based on shared regional characteristics. In sum, the primary objective of this chapter is to provide a generic research agenda concerning border regions in Central America.

INTEGRATION OR DISINTEGRATION: TOWARDS A CONCEPTUALIZATION OF BORDER REGIONS IN CENTRAL AMERICA

The development of a research programme of relevance to Central America as a whole cannot be undertaken without taking into account the characteristics of the economic and political crises the region has undergone over the last decade as well as their particular impact on the dynamics of border regions. In this sense, even though the nature of the crises has varied from one country to the next, with differing impacts on bi- and multilateral relations, the reality of the past twelve years in Central America has fomented a context of reduced possibilities for the region's economic and political integration. The regional perspective has been progressively replaced by processes of economic and political disintegration which have strengthened the idea of national frameworks as opposed to Central American frameworks. This last aspect has been paralleled by a relative process of de-nationalization due to the impact of political and economic interventions in favour of structural adjustments, privatization and a weakening of the State. These initiatives have been orchestrated chiefly by institutions from outside the region such as the international banking system, the International Monetary Fund, and pressures exerted by the US Government, etc.

During the last decade, two distinct and opposing contexts could be identified in terms of potential future developments in Central American countries. In 1986, Lungo, in discussing the situation of border regions in the isthmus, identified these two possible contexts during the turbulent decade of the 1980s. The first would be the possibility of direct outside intervention in the region, and an ensuing generalized war, in which case the joint development of border regions would have been relegated to a remote priority for the countries of the isthmus. With the outbreak of war, border regions were to become sensitive areas for military authorities and the only planning initiatives to affect them would have been military contingency plans. Clearly, academic research on border regions would have been unlikely under such circumstances.

The alternative context would have been characterized by the rise of pluralistic, self-determined and pacific governments in each of the region's countries. In this situation, the concerted development of border regions would have played an important role in the consolidation of a model of integration for Central America. The widespread pessimism of the mid- to late-1980s concerning the real possibility of peace in the region has given way to a more positive outlook since the early-1990s. The dramatic political evolution which has occurred since the electoral defeat of the Sandinistas in Nicaragua, and the signing of the Peace Accords in El Salvador have largely contributed to the change of atmosphere in the region. Under these circumstances, there is an urgent need for a comprehensive research programme on border regions in Central America, given the renewed interest and debates over the possibilities of future economic or political integration in the region. Seen from this perspective, one of the major challenges to Central American researchers in the social sciences is to contribute to the construction of a new regionality to replace the now obsolete concept of economic integration weakened by the crises of the 1980s. This new regional perspective must offer a solid conceptual basis for the resurgence of integration efforts in the region.

It is on this basis that arguments are constructed in terms of generic types of research which need to be promoted in the future. The proposed research agenda for the region is based on a characterization of the economic, social and political conditions prevailing in Central America's border regions, which are a product of past and present conjunctures.

The economic and political crises of the past decade have modified in many ways the forms of social organization, formal boundary relationships and the articulation of borderland economies to the national context as compared to the decade of the 1970s. For example, the massive exodus of refugees fleeing war zones in Nicaragua and El Salvador and their present-day return has dramatically affected border regions throughout Central America. Consequently, detailed knowledge of these new regional realities is important prior to promoting schemes of bilateral cooperation and regional integration. In order to provide an instrumental conceptualization of the types of border regions existing in Central America, the following typology is proposed:

(a) Border regions of tension–contention

In this type of border region, considerations of a geopolitical and military nature have dominated border relations in the past, due either to the ideological confrontation between neighbouring countries, or to the impact of civil war and low-intensity warfare across borders often used as refuges for paramilitary groups (Granados and Quesada 1986). This category includes the border regions of Nicaragua and Costa Rica, Nicaragua and Honduras, El Salvador and Honduras and, to a lesser extent, of Guatemala and Belize. Despite the fact that the geopolitical and military tensions which marked these regions during the 1980s have largely subsided, the impact of these conflicts on social formations, regional trade networks and natural resource bases has left long-lasting structural marks in these border regions. This fact conditions in many respects these regions' potential for integration in the future.

(b) Transnationalized border regions

Here, economic activities on one or both sides of the border are controlled by foreign transnational companies, and cordial and pacific relations generally prevail between the neighbouring states. This category of border region can be found between Costa Rica and Panama and between Honduras and Guatemala. In the case of the Costa Rica–Panama border region, banana-producing enclaves function as transboundary operations, particularly along the Sixaola–Changuinola–Almirante axis, and to a lesser degree along the Golfito–Ciudad Neily–Puerto Armueles line (see (c) below (Asymmetrical border regions)). Moreover, this type of border also exists between Belize and Mexico, where sugarcane production controlled by multinational companies predominates.

(c) Asymmetrical border regions

In this type of border region, differing social and economic conditions exist on either side of the boundary, with clear implications of structural inequality (that is to say greater or lesser degrees of development, salary differentials, varying availability of social services, etc.). In these asymmetrical border regions, there have been sharp variations in the exchange rates of national currencies and in the local price structures of consumer goods and production costs, due to the impact of the economic crises suffered during the 1980s and continuing in the 1990s.

The clearest example of this type of border is that between Costa Rica and Panama, on the Pacific Coast, where higher wages in Panama attract Costa Rican labourers, while better health care and education in Costa Rica provide incentives for transboundary linkages between both sides. Similar border situations can be found between Belize and Mexico, and between Guatemala and Mexico.

(d) Refugee border regions

This category includes those areas which underwent processes of occupation by refugee migrants fleeing from the economic, social and political conditions prevailing in their own countries, particularly during the 1980s. This type of border region is best illustrated by the interior region of Belize bordering on Guatemala where, in a sparsely inhabited area, uncounted numbers of Guatemalan and Salvadorean peasant and indigenous refugees settled and created their own local economy under conditions of complete tolerance on the part of the Belizean authorities. On the other hand, in border regions of tension–contention, the presence of refugee camps is subject to varying levels of military control and acceptance by local authorities. Such is the case of the southern Mexican border with Guatemala, in the area of the State of Chiapas (see Chapter 5) and the Department of Huehuetenango and San Marcos in Guatemala (Castillo 1986).

(e) Overlapping ethnic–cultural border regions

In this case, national borders 'are diffuse ... creating informal territories in which cultural and ethnic landscapes overlap' (Morales 1986). This category is typified by the border region between Nicaragua and Honduras, particularly along the Coco River and the surrounding areas traditionally settled by Miskito indians, which effectively span the boundary. Similar situations of transboundary ethnic groups are found between Costa Rica and Panama, where Guaymi and Bribri culture groups occupy both sides of the border; also Mexico and Guatemala, Honduras and Guatemala and, in the case of Panama, the Darien Gap bordering on Colombia, all feature transboundary culture groups. These ethnic–cultural regions are frequently combined with other categories of border regions previously described. Unfortunately, many regions where pre-existing culture groups spanned borders were the setting for armed conflicts, refugee settlements, transnational economies and asymmetrical relationships. These culture groups overlapping across national

53

boundaries often became the object of repression by armed forces engaged in low-intensity warfare, as in the case of the Nicaragua–Honduras border, where the Miskito people led a daily struggle to maintain transboundary linkages with their kin.

In order to complete a conceptualization of the main types of border regions existing in the Central American isthmus, it is necessary to mention a particular context which, while not in itself constituting a 'border region', does in fact constitute a logical extension of this idea, as a zone of contact and interaction between neighbouring nation-states. The problem of maritime boundaries in Central America indeed constitutes an issue which cannot be overlooked, as Sandner (1987) has well pointed out. Central America's most conflictive maritime border areas are the Gulf of Fonseca, where the boundaries of El Salvador, Honduras and Nicaragua converge, and the Gulf of Honduras, where Belize, Guatemala and Honduras' boundaries merge and even overlap. Both these areas exhibit particular features as coastal border regions and, while the first has been characterized by latent tension and conflict, the latter is marked by relatively peaceful relations where coastal trade and exchange predominates.

In summary, the conceptualization of the types of existing border regions in Central America here proposed offers an adequate thematic framework in which to address future research priorities. Three broad thematic elements and the type of border region they condition can be distinguished:

1   Geopolitical and military elements, which largely condition border regions of tension–contention;
2   Elements of economy – production, consumption, flow of consumer goods and labour force – which are key factors in asymmetrical and transnationalized border regions;
3   Human and cultural elements and migration, flows which particularly affect refugee and ethnic border regions.

These thematic categories will serve as a basis for the identification of potential research priorities on border regions in Central America, as will be discussed in the third section of this chapter.

TOWARDS A RESEARCH AGENDA ON BORDER REGIONS IN CENTRAL AMERICA

Tension–contention border regions

Three basic types of research can be identified concerning this particular category of border regions.

1   Research aimed at analysing the physical, social, economic and political changes brought about in these border regions over the past decade, and their role in promoting or impeding future integration efforts between neighbouring countries. For instance, these changes can include pressure on natural resources linked to processes of population movement and settlement, destruction of infrastructure and services due to ideological and ethnic conflicts and the strengthening of nationalist attitudes, etc. This type of analysis must centre on the way in which conditions imposed by different political situations modify the overall structure of the border region in physical, economic and social terms.

2   Research focusing on state policies (including large infrastructure projects) during a given period, in order to analyse their bearing on binational and multinational integration or disintegration initiatives. Further, this approach may shed light on the processes of territorial consolidation and integration at a national level. In this type of research, it is important to consider border regions from a dual perspective: first, as areas of separation and contact between two sovereign states and, second, as peripheral regions of single nation-states, that is to say in terms of their internal structural function within a given national territory.

3   Research addressing the forced migration of refugees into border regions of host countries, their social and material living conditions and their impact on local economies and societies, including the problems associated with health and sanitation control policies.

Transnational border regions

1   Research focusing on the existing forms of economic organization and trade, taking into consideration production systems, formal and informal trade, and labour force mobility from one side of the border to the other.

2   National differences in terms of social and material living conditions.

3 Conflicts and contradictions at a local and regional level between the power and influence of transnational corporations and local, regional and national governments.
4 Aspects related to the national and cultural identity of borderland populations.

Asymmetrical border regions

1 Research addressing historical changes in patterns of transboundary trade and local production systems, due to changing national economic conditions (such as fluctuations in currency exchange rates and relative devaluation of national currencies), economic crises and their local impact over the past decade and a half. The response to these fluctuations by formal and informal economies, including the spatial patterns of contraband, constitutes an important item to address.
2 Patterns of organization and changes in the transboundary labour market, considering both seasonal and permanent fluctuations.
3 Public policies and the regional integration of different and asymmetrical economies.
4 Regional and national identities.

Refugee border regions

1 Forms of population and territorial occupation, and the impact of settlement patterns and schemes on existing ecosystems and agricultural production.
2 Patterns of local and regional cultural integration.
3 Impact on local and regional markets, both seasonal and permanent.
4 The social and material living conditions of refugees, and their impact on local health conditions and disease control policies.
5 The analysis of state policies towards migrant-refugee groups.

Ethnic–cultural border regions

1 Research should address the changes in lifestyle and subsistence economy experienced by ethnic groups as a result of the interference of external agents (be they economic, political, military, etc.). These impacts may include settlement changes, the alteration of traditional cultural-ecological resource-use patterns, modifica-

tion in subsistence agriculture systems and employment, as well as grassroots organizations and local power structures.

Finally, in terms of maritime border regions such as those of the Gulf of Fonseca and the Gulf of Honduras, the future research agenda should include considerations of geopolitical and legal aspects, and human and commercial flows in areas previously marked by conflicts and tensions. Second, research should be geared towards the reliable delimitation of maritime boundaries and the regulation of the transboundary flow of goods and persons (linked to labour markets).

CONCLUSION

This chapter offers a synopsis of border regions in Central America. While not providing an exhaustive analysis of the problems associated with boundaries in the region, this contribution offers a synthetic appraisal of the types of borderlands encountered in the isthmus. The typology of border regions proposed constitutes, in a sense, a first approximation of the problem, and should serve to guide further research. It is hoped that this agenda for future research will stimulate social scientists and lawmakers to focus on the problems inherent in the border regions of the Central American region.

BIBLIOGRAPHY

Castillo, M.A. (1986) 'Algunas determinantes y principales transformaciones recientes de la migracion Guatemalteca a la frontera sur de Mexico', *Estudios Sociales Centroamericanos*, No. 40.

Granados, C. and Quesada, L. (1986) 'Los intereses geopolíticos y el desarrollo de la zona nor-atlántica costarricense', *Estudios Sociales Centroamericanos*, No. 40.

Ireland, G. (1941) *Boundaries, Possessions and Conflicts in Central and North America and the Caribbean*, Cambridge: Harvard University Press.

Lungo, M. (1986) 'Una panorama histórico de las regiones fronterizas en Centroamerica: Seis tesis y dos Hipótesis', *Estudios Sociales Centroamericanos*, No. 40.

Morales, M. (1986) 'Crisis del estado nacional y los problemas territoriales fronterizos en Centroamerica', *Estudios Sociales Centroamericanos*, No. 40.

Sandner, G. (1981) 'La planificación regional integrada como agente del estado frente a la comunidad local y la patria Chica: Un resumen de experiencias Centroamericanas', International Geographic Union, *Seminar on Regional Development Alternatives in the Third World*, Belo Horizonte, Brazil: International Geographic Union.

——— (1987) 'Aspectos de problemas de geografía "territorial" del Mar Caribe en el contexto de las nuevas delimitaciones', *Geoistmo*, 1, 1:9–32.

5

CHIAPAS, CENTRAL AMERICAN FRONTIER IN MEXICO

Jan de Vos

INTRODUCTION

The State of Chiapas does not seem at first sight to belong to the area commonly identified as Central America, for since 1824 it has formed part of the Mexican Republic. Rather it constitutes, together with Tabasco, Campeche and Quintana Roo, the region where Mexico borders on Central America. Hence the name 'southern border' which Mexican politicians and academics have been using for the area for the past decade. It is, obviously, a characterization created from the centre of the country. The inhabitants of the four states that border on Belize and Guatemala have never felt themselves to be 'frontiersmen' in the sense of being alien or opposed to the Central American complex. Rather they have identified themselves by the term 'Mexican south-west', used generally before the introduction of the new term. The feeling is easily understandable on account of the increasingly greater integration of these four states into the national process since independence. For analogous reasons, the indigenous peoples of the Cuchumatanes region, for instance, also prefer to consider themselves inhabitants of Guatemala's north-east rather than inhabitants of its northern border.

These considerations, regardless of how valid they may seem, lose ground before the current fashion which insists that we think and speak in terms of *frontiers*. I take part in the discussion with the present contribution, whose title attempts to come to terms with the central theme, but with the idea of contributing an alternative point of view. I could have chosen to look upon Chiapas as the frontier of Mexico with Central America, which would have been the more obvious stance, at least from the Mexican perspective, as well as the easier, since I would have had to take into account only recent history. However, I decided to shift to the Central American point of view and to go back a few centuries into the past.

For those of us who study the colonial period, the very particular position of Chiapas among the Mexican border states is worthy of note. For almost three hundred years its territory was administered from the city of Guatemala, seat of the *Audiencia* district of the same name. The Chiapans thus shared, for almost three centuries, the same political, economic and sociocultural destiny as the inhabitants of Costa Rica, Nicaragua, Honduras, El Salvador and Guatemala. It is thus not strange that, even today, Chiapas continues to be, in several important aspects of its personality, a land with a 'Central American flavour'. Hence the thesis stated in the title of this chapter: Chiapas, Central American frontier, not so much with, but rather in, Mexico.

The Central American vocation of Chiapan society is the product of a long process which began in 1528 with the coming of Spanish *conquistadores* and Indians from Guatemala, and with the incorporation, in 1531, of the province of Chiapa into the government of Pedro de Alvarado. The system of *Audiencias*, introduced into Central America in 1544, consolidated this state of affairs because not only Chiapa but also the coastal area of the Soconusco became dependencies of the headquarters in Guatemala (see Figure 5.1). The adherence to Mexico, first of Chiapa (in 1824) and later of Soconusco (in 1842), did not mean for either of these two regions a radical negation of their colonial inheritance. Nineteenth-century Chiapan society, politically more and more separate, continued to be Guatemalan in speech, in way of life, in religious practices, in artistic tastes, in modes of agricultural production and cattle raising, in racial divisions, in social antagonisms, and so on. This fraternity has been maintained right up to the present, despite the different roads taken by Mexico and Guatemala after the revolution of 1910.

It is worthwhile analysing the degree to which Chiapan society became involved in the Central American sociocultural process. At the same time, it would be interesting to study the phenomenon in its most diverse expressions, from eating habits to religious and ethical convictions. This is a little-explored area which has yet to attract the attention of anthropologists, sociologists or even psychologists. In the present instance, I shall restrict myself to sketching the Central American orientation of Chiapas with regard to the course of its history. As I see it, this begins with the establishment of the colonial system in the isthmus. It was the Spanish who introduced, mainly for administrative

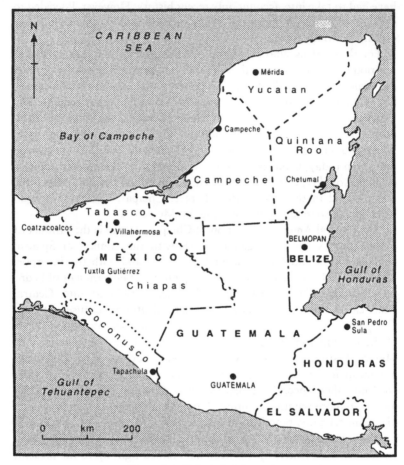

Figure 5.1 Chiapas border region

reasons, the division between Mexico and Central America. It is they who are responsible for the two sociocultural complexes, each of which developed its own personality. With independence, these two personalities acquired different destinies: the Mexican became ever more centred – not to say centralized – while the Central American, on the other hand, became more fragmented – not to say torn apart.

In this world, then, divided in two spatially and temporally, Chiapas found itself in a frontier situation *par excellence*, changing from a Central American province into a Mexican state. The key moment in this process was the year 1824. It divides Chiapan history in two, into a before and an after. This cleavage becomes dominant when one tries to analyse the implications of that change. I shall follow it in the overview that I shall present here. I shall expound how the Chiapans received their Central American vocation; how they came to deny it, giving to their sociopolitical life a completely opposite course; and how they remained, in spite of this change, a people 'with a Central American flavour'. I shall pay special attention to the decisions which influenced the forming of that 'flavour' and to the motives behind those decisions. Thus I shall review: 1) the geopolitical ambitions of the *conquistadores*; 2) the administrative concerns of the Spanish Crown; 3) the will to power of the local oligarchies; 4) the reasons of state of the national government; 5) the financial interests of the German farmers; 6) the labour aspirations of migrant workers; and 7) the mortal anguish of political refugees. The first two motives gave origin to the Central American inheritance; the next two explain its abandonment; the last three explain its persistence in spite of the rupture.

GEOPOLITICAL AMBITIONS

It is a fact that the majority of the Central American states and their neighbours of the Mexican south-west came into being out of the struggles between different *conquistadores*. Nor is there any doubt as to the motive that dominated this clash of interests: the ambition of each to take to himself as much territory as possible with the ultimate hope of converting this into a semi-independent principality. Chiapa and Soconusco – the two colonial provinces which in the nineteenth century made up the State of Chiapas – were coveted by at least four candidates: Hernán Cortés, Pedro de Alvarado, Francisco de Montejo and Alonso de Estrada. The first lost no time in assigning to himself the rich province of Soconusco as a personal *encomienda*. The measure had little effect on the inhabitants, since they simply changed owners, having for almost half a century paid tribute to the Aztec ruler. This orientation of Soconusco toward Mexico, imposed in preHispanic times, suffered no change when Cortés lost his rights in 1528. In that year, the province was declared an *encomienda* of the Crown but continued to depend administratively upon the *Audiencia* of Mexico which had just succeeded Cortés in the government of New Spain.

In contrast to Soconusco, which for a decade was the exclusive property of Hernán Cortés, the province of Chiapa became the apple of discord, fought over by three rivals. It was militarily conquered by Mexico in 1524 and 1526 but was never really pacified; in 1528 it was invaded again by two contingents of Spanish soldiers: one from Mexico under the command of Diego de Mazariegos, Lieutenant of the Treasurer Alonso de Estrada, then governor of New Spain; and one from Guatemala, commanded by Pedro Puertocarrero, Lieutenant of Pedro de Alvarado, since December of 1527 the *de jure* governor of Guatemala and Chiapa. The clash between the two captains was inevitable; Mazariegos prevailed, being then the stronger. Although he had founded a town near the Indian village of Comitán, Puertocarrero was forced to retire to Guatemala leaving the territory won in the hands of his adversary, with the result that the province seemed definitely to enter the Mexican sphere. However, three years later Pedro de Alvarado, returning from Spain, was able to impose on the *Audiencia* of Mexico the royal order of 1527. For the next ten years Chiapa was governed from Guatemala until 1539, when Alvarado came to terms with Francisco de Montejo and exchanged the province for Honduras. Thus Chiapa became a part, along with Tabasco and Yucatan, of the vast Mayan territory which Montejo was about to create for himself and which he hoped to pass on to his descendants.

ADMINISTRATIVE CONCERNS

The ambitions of Montejo and his colleagues came to an end in 1544 with the establishment of the *Audiencia de los Confines*, first in Gracias a Dios, later in Guatemala. Created sixteen years after that of Mexico, the new governmental institution received jurisdiction over 'the provinces of Guatemala and Nicaragua, Chiapa, Yucatán and Cocumel and Higueras and Cabo de Honduras and whatever other provinces and islands there may be ... as far as, and including, the province of Tierra Firme called Castilla de Oro'. A few years later, the Crown took steps of great import for the future of the region. In 1556 it separated the province of Soconusco from the district of the *Audiencia* of Mexico and placed it under the jurisdiction of that of Guatemala. In 1560 it decided just the opposite for Tabasco, Yucatán and Cozumel. These three provinces passed out of the Central American sphere to form part of the administrative territory of the *Audiencia* of Mexico. The provisions of 1556 and 1560, together with the royal order of 1543, established the base of a territorial, political and sociocultural definition which soon received the name of 'Kingdom of Guatemala'.

Chiapa and Soconusco remained within this kingdom until its dissolution in 1821. For more than two-and-a-half centuries, their societies grew and diversified under the influence of the governing centre of the city of Santiago de los Caballeros. To this capital were sent, from Ciudad Real in Chiapa and Huehuetán in Soconusco, the tribute of the Indians, the products of the *mestizo* artisans, the intellectual hopes of the creole youth, the complaints of the holders of *encomiendas* and the quarrels of the local chieftans. From there came life styles, artistic tastes, royal provisions, the visits of judges and officers. It was this subordination to the Guatemalan administrative capital that, little by little, brought Central American elements to the formative process of the Chiapan and Soconuscan idiosyncrasies.

These two units would have continued their separate development had it not been, again, for the intervention of the Crown. In 1786 the colonial province of Chiapa, since 1769 divided into two major *alcaldías* – Ciudad Real and Tuxtla – was incorporated into the government of Soconusco to form a new fiscal and administrative unit, the *Intendencia General de Ciudad Real*. With this measure, the way was open for the Soconuscans to make common cause with the Chiapans with whom, up to that time, they had been united only in the ecclesiastical sphere.

If we examine the motives behind the decisions of the colonial government we find, from the beginning, a concern on the part of the Crown to ensure control over its overseas possessions. The creation of the *Audiencia de los Confines* answered to the wish to eliminate in Central America the possible rise of semi-independent principalities along the lines of the *Marquisate* of the Oaxaca Valley. Later administrative changes, such as the incorporation of Chiapa and Soconusco and the separation of Tabasco and Yucatán, seem to have been part of the same strategy: to continue dismembering the territory put together by Montejo. The measure of 1786 at last becomes part of the general policy of the Crown aimed at putting an end to the abusive power of the major *alcaldes* and the traditional governors. The new system of *intendencias* was established with the aim of centralizing the administration and streamlining the tax system.

WISH FOR POWER

In 1821, the year of independence from Spain and incorporation into the Mexican empire, Chiapans and Soconuscans had but thirty-five years of common existence under the same *intendencia*. On the other hand, their links to the Kingdom of Guatemala went back, for both, to the middle of the sixteenth century. The power vacuum created by the disappearance of the colonial regime and, a little later, by the overthrow of Iturbide was ably taken advantage of by the creole oligarchies. These then controlled the cities of Comitán, Chiapa and Ciudad Real, and the towns of Tuxtla and Tapachula. These small but influential groups of local notables were the ones that, from 1821 to 1824, made the decisions on the future of the new state born of the Bourbon *intendencia*. The rest of the population – a large mass of Indian peasants and *mestizo* artisans – was not invited to participate in the affair in spite of the democratic appearances of the open *cabildos* or general juntas which then took place in all the municipalities.

The process of the decisions taken in those three memorable years can only be explained if we understand the motives behind them. The oligarchy of Ciudad Real was influenced by its aversion to the bureaucrats and businessmen of Guatemala and by its hope of receiving more political posts and economic advantages from the Mexican Government, even as this was changing from imperial to republican. On the other hand, the elites of Tuxtla, Chiapa and Tapachula seriously considered, in 1823, the possibility of adding Chiapas as a sovereign state to the recently created Central American Federation. In 1824, the polarization of interest reached such a degree as to lead the oligarchy of San Cristóbal to manipulate the plebiscite in order to win the elections. At the same time, the powerful families of Tapachula took the no less serious decision to separate the district of Soconusco from the rest of Chiapas and hand it over to the Central American Government. In both cases a few imposed their will on the rest and led the entire body in opposite directions because it suited their class interests. This is worth emphasis, since official historiography has created several myths around the incorporation of Chiapas into the Mexican Republic and around the separatist movement of Tapachula. Unfortunately, these myths have acquired currency over the years, to such an extent that they are accepted today as historical truth.

The debate over the electoral fraud during the plebiscite of 1824 will no doubt continue to be heated for some time to come. What is beyond doubt is the turn that Chiapan society then took, without the Soconuscans up until 1842, and including them afterwards. In both cases, Chiapas turned its back on the Central American sphere and turned its glance toward the Mexican capital. The new experience would have its ups and downs since Chiapas continued to be a forgotten region on the fringes of the national process. In some measure, these reverses were the result of its frontier position, now more than one thousand kilometres from the Federal District, the place where decisions were made.

Such would be the isolation – and the aversion arising from it – that

64

in 1914, with the arrival in their territories of the army of Carranza, the oligarchies of the time set up their own successful counter-revolution. But this later process does not take away from their ancestors of 1824 the merit of having themselves – and not some distant, powerful authority – programmed their own future.

REASONS OF STATE

The decision of the Chiapans to get out of the Central American sphere was curiously interpreted by the government of that federation as vile theft on the part of Mexico. On the collapse of the federation in 1839, the leaders of the new state of Guatemala came to an even stranger conclusion. Without taking into account the fact that the new Guatemala, as an independent country, was but a fraction of the now defunct Central American Union, it believed it had the right to demand of Mexico the return of Chiapa and Soconusco, as if they had been a part not only of the Kingdom of Guatemala but also of the older colonial province of Guatemala. This unfortunate identification was the work of the Guatemalan governing circles, the politicians as well as their partisan intellectuals, who were tireless in taking their inflammatory message to the people through innumerable books, articles, pamphlets and speeches. The frustration of the ruling class thus truly became a collective trauma, since the citizenry ended up believing that it had been deprived by Mexican arrogance of something that was really theirs.

In the face of this orchestrated resentment, the Mexican Government could find no adequate response. This is clearly demonstrated by the subsequent history of diplomatic relations between the two countries. Exactly fifty years were to pass, between 1824 and 1874, before the first meeting took place between the two statesmen who would eight years later achieve the solution of the conflict through a border treaty. It was not accidental that these two protagonists, Matías Romero and Justo Rufino Barrios, possessed farms located along the border strip, and that the meeting took place at El Malacate, the Soconuscan hacienda of Barrios. It was in that place and at that moment that the government of Guatemala, in the person of President Barrios, began for the first time to adopt reasons of state instead of visceral attitudes. President Barrios was dreaming then of reviving the Central American Federation, with himself at the head, and he was willing to put an end to the Chiapan syndrome in exchange for a definitive agreement on the border with Mexico. He sought US arbitration as an honourable way out of so many years of international turmoil that grew out of the preposterous pretensions of his predecessors. The result is well known to all. The treaty was signed in the city of Mexico on September 27, 1882. In the first article of the treaty 'the Republic of Guatemala forever renounced the rights it believed it had to the territory of the State of Chiapas and its District of Soconusco, and consequently considered the said territory as an integral part of the United States of Mexico'.

The victory of reason over resentment was brief. No sooner was the treaty signed and ratified than the frictions of the past reappeared. A first sign was the obstruction on the part of the Guatemalan authorities of the actual drawing of the dividing line. With the final agreement of 1895 the problem which heretofore had been both territorial and mental was reduced to the latter, where it took on new vigour. It continues alive to this day, if we are to believe the assessment of Luis Zorrilla, the ultimate scholar and greatest expert on the matter. According to Zorrilla,

> the Guatemalan feeling of ownership persists, as does the feeling of being the victim of despoilment and perpetual aggression by Mexico; thus, during the Mexican Revolution, official claims reappeared, and to the present day there are educated and distinguished persons who ... simply turn a deaf ear on any reasoning that can weaken their case. This blindness is general: governmental, political, diplomatic, journalistic, pedagogical and historical.
>
> (Zorrilla 1984)

MERCANTILE INTERESTS

The fixing of the international frontier in 1882 almost coincided with the promulgation, in 1883, of the Mexican Government's Colonization Law. In that same year, the government entered into contracts with surveying companies, charging them with surveying and selling to colonists, preferably European, the immense tracts of virgin land that still existed in the country. In the Soconusco this task was given to the Mexican Land and Colonization Company whose president was the Englishman Louis Huller. By means of three successive agreements, this company acquired an area of more than a million and a half hectares, among them almost the totality of lands suitable for growing coffee. The purchasers were in large measure German farmers who, two decades before, had begun cultivating coffee in the neighbouring Costa Cuca of

66

Guatemala and who were interested in extending coffee growing to the western slopes of the Sierra Madre of Chiapas.

The arrival of the German-Guatemalan coffee growers was a silent but effective invasion. In less than ten years, the growers, whose names even today dominate the coffee business in terms of capital and land holdings, had established themselves in Mexico: Lüttmann, Giesemann, Edelmann, Pohlenz and Kahle, among others. In the first decade of the twentieth century, the number of German farms grew considerably, to the detriment of the holdings of Mexicans, North Americans, English, French and Swiss. By 1909 three-quarters of the Soconuscan coffee plantations were in the hands of German owners or managers. These growers, few in number but economically powerful, received financial and organizational support from the large mercantile and banking companies of Hamburg. It was these Hanseatic firms which were the true promoters and beneficiaries of the upsurge that intensive coffee growing enjoyed at that time in the Soconusco. They provided the region with a modern imported infrastructure and they opened production to the international market. They also set the conditions for the development of labour relations and, significantly, the greatest income to the Chiapan state treasury came not from the tax on the commercialization of coffee but from the enlisting of indigenous labourers from the Altos de Chiapas.

There is no doubt that the German farmers of the Soconusco dominated, for almost half a century, the shameful traffic in labour. They also managed to influence indirectly the legislation enacted during that same period by the Chiapan Government justifying those labour practices. The farmers thus received from the authorities in Tuxtla extralegal room to act as they wished. The price of that privilege was the handing over to the State Treasury of a healthy annual sum. The farmers lost the monopoly in 1936 when the government itself assumed control through the Department of Social Action, Culture and Protection of the Indigenous, and the Indigenous Workers Union. Comparing that whole process with the *mandamiento* system then operating in neighbouring Guatemala, the relationship between the two is obvious. And this similarity is clearly traceable to the origin of its instigators, the German-Guatemalans who apparently imported not only their persons but the practices of exploitation as well.

LABOUR ILLUSIONS

The Tzotzils and the Tzeltals of the Altos de Chiapas were not the only contingent of labourers who yearly came down to the Soconuscan farms; they were also attractive to the peasants of west Guatemala. Many of them crossed the newly drawn frontier to evade the severe effects of the Labour Code instituted in 1876 by the liberal government of their country. In Guatemala, any person not a proprietor of a certain specified extent of land was obligated by law to sign a labour contract. And the majority of Guatemalan peasants fell into this category. Furthermore, the wages in the western Departments were much lower than in neighbouring Soconusco: from one to two *reals* as against four or six *reals*, depending on whether the person was free of debt or a debtor. Thus it was bad working conditions that impelled thousands of Guatemalan Indians toward emigration.

The majority of these migrants were temporary, being employed during the months of the harvest and the clearing of the plantations. Nevertheless, there was a group that established itself permanently in the border area of the Chiapan Department of Mariscal, a region that became part of the Republic of Mexico by the border treaty. The towns and villages that constituted it had been Guatemalan and, despite their naturalization as Mexican, they continued as Guatemalan for the immigrants. These came to share with their former countrymen the difficult living conditions brought about by the poverty of the land and the despotism of the local authorities. They also shared the hope inspired after 1914 by the presence of constitutionalist troops in Chiapas. It was the peasants of Mariscal who were first to become aware of their ability to organize themselves politically and in unions and thus to enter into negotiations with the farmers of the Soconusco. And there is still research to be done on the participation of these Guatemalans – immigrant and naturalized – in the founding of the Socialist Party of Chiapas, on January 15, 1920. The location of that transcendent event was Motozintla, the seat of the Department of Mariscal and, before 1882, a Guatemalan town belonging to the Department of Huehuetenango. It is noteworthy that, in the founding document, the founders of the party included as a future programme of action 'the disappearance of frontiers, these odious lines drawn by the egoism of our ancestors'.

To get an idea of the magnitude of the migratory flow of Guatemalan peasants toward Chiapas at that time, the information provided by Erasto Urbina, from 1935 to 1944, an official of the Union of Indigenous Workers of Chiapas, is helpful. It was calculated that at the time in the coffee plantations of the Soconusco there were some five or six thousand settled peons, almost all of them Guatemalan, and between thirty and forty thousand temporary labourers, of whom ten thousand came from the Altos de Chiapas and the rest from Guatemala. The first were integrated into the peasant class of the region through the agrarian settlement in the 1940s, while the others, or rather their sons and grandsons, return each year to the same farms where their fathers and grandfathers sought an alternative to the forced labour in their country of origin. Since 1960 the Guatemalan day labourers have come in increasing numbers to offer their labour, until they became the principal source of labour. In 1978, their number was calculated at 32,000. Four years later, in 1982, they were already 75,000, a number which can be explained only by the pressure of the genocide begun in Guatemala the previous year and by the consequent willingness of the exiles to work for minimum wages, thus displacing the Chiapans.

MORTAL ANGUISH

The Guatemalan peasants who, since 1900, have crossed the Suchiate River in search of work on the Soconuscan farms are considered 'economic migrants', a term today much in vogue. On the other hand, the waves of families which, beginning in 1981, flooded the Mexican side of the frontier strip along all its width, are called 'political refugees'. The two definitions are generally accepted and, in our case, they are happily operative since they permit the identification of the Guatemalan refugees as a new Central American presence in Chiapas, different from that of the migratory labourers. The latter can prepare their trip, explore the labour market, carry resources with them and return home if the venture does not take off. The former, on the other hand, are people fleeing their country under emergency conditions, impelled by the fear of persecution or by actual persecution.

In 1983, the population of Guatemalan refugees was to be found in seventy-seven settlements, sixty of them villages, the remainder camps. The total number of persons, according to the estimates of UNHCR–COMAR, came to 38,000. Of course, there were several thousand more who could not be counted. These are the ones who were integrated into Chiapan communities, worked on the farms of the Soconusco or were living hidden among the urban masses of Tapachula and Tuxtla. A figure closer to reality would be some 50,000 persons. Almost all are Indian peasants who crossed a border which to them was artificial in many ways. The majority are speakers of some Mayan language and inhabitants of the border strip. Because of this condition of racial, cultural and territorial proximity, they maintain, from some time back, close ties of commercial and personal exchange with their neighbours on the Chiapan side.

The refugee population is notably asymmetrical in sex and age. Women and young people predominate, since the flight took its toll of lives among the elderly, and the adult males were assassinated or remained in their place of origin to fight in the insurgency, or to take care of the crops or herds. Another characteristic is the resistance of the refugees to becoming 'westernized'. Even outside their natural environment they try to retain their communal structures. The trip to Mexico is often undertaken in family groups and in communities of origin. Upon arrival, they reproduce as soon as possible the communal organization they had in their village of origin. They also keep alive the hope of one day returning to their homes and corn fields, since many of them had small holdings in Guatemala. This tie makes them reluctant to go very far from the dividing line.

Chiapan society has reacted in different ways to the arrival of the refugees. At first there was a general attitude of openness, expressed in the provision of housing and pity for their plight. In time, many good intentions gave way to resentment, especially on the part of peasants equally hard hit by the economic crisis affecting their own country. They often end up feeling threatened by the competition for resources as basic as fuel, water, work and land. There is also always some possibility of jealousy on account of the free and prolonged aid given to the refugees by international organizations, both governmental and voluntary. Another attitude, also rather contradictory, is that of taking advantage of the abundant labour force available, lowering salaries and increasing the cost of food. In this context, the farmers of the Soconusco of course consider the refugees a blessing.

CONCLUDING REMARKS

With the anguish of the refugees we conclude this brief review of the motives and actions that brought about and then formed the Central American vocation of Chiapan society. We saw how the ambitions of the *conquistadors* and the interventions of the Crown brought the Chiapans and the Soconuscans into the Kingdom of Guatemala, how the interests of the local oligarchies made them change direction, and how the desires of the farmers, the labourers and the refugees from Guatemala gave new life to the original orientation. All of these motivations – and the decisions born of them – explain how Chiapas came to be, in some way, a 'Central American frontier in Mexico'.

BIBLIOGRAPHY

Aguayo, S. (1985) *El éxodo centroamericano. Consecuencias de un conflicto*, Mexico City: Secretaría de Educación Pública.

Baumann, F. (1983) 'Terratenientes, campesinos y la expansión de la agricultura capitalista en Chiapas, 1896–1916', *Mesoamérica*, Cuaderno 5 , June: 8–63, Guatemala: CIRMA.

Benjamin, T. (1989) *A Rich Land, a Poor People. Politics and Society in Modern Chiapas*, Albuquerque, NM: University of New Mexico Press.

——— (1990) *El camino a Leviatán. Chiapas y el Estado Mexicano, 1891– 1947*, Mexico City: Consejo Nacional para la Cultura y las Artes.

Cosio Villegas, D. (1960) *Historia Moderna de México. El Porfiariato. La vidapolítica exterior*, Mexico City: Editorial Hermes.

De Vos, J. (1988) 'El sentimiento chiapaneco. Cuarteto para piano y cuerdas. Opus 1821–1824', *Revista ICACH*, tercera época, Tuxtla Gutiérrez, Chiapas, 3: 30–50.

——— (1991) *Los enredos de Remesal. Ensayo sobre la conquista de Chiapas*, Mexico City: Consejo Nacional para la Cultura y las Artes, Serie 'Regiones'.

Fenner, J. (1986) *Lebens – und Arbeitssituation der indianischen Kaffeeplantagenarbeiter in Soconusco, Chiapas*, Ph.D. thesis, Univesity of Hamburg.

Garcia De Leon, A. (1985) *Resistencia y Utopía. Memorial de agravios y crónica de revueltas y profecías acaecidas en la provincia de Chiapas durante los últimos quinientos años de su historia*, Mexico City: Ediciones Era.

Gerhard, P. (1979) *The Southeast Frontier of New Spain*, Princeton: Princeton University Press.

Spencer, D. (1989) *El Partido Socialista Chiapaneco. Rescate y reconstrucción de su historia*, Mexico City: Ediciones de la Casa Chata–CIESAS.

Zorrilla, L.G. (1984) *Relaciones de México con la República de Centro America y con Guatemala*, Mexico City:Editorial Porrúa.

6

THE MEXICO–BELIZE BORDER

Origin, present situation and perspectives

Alfredo C. Dachary

THE ORIGINS

The Mexico–Belize border was legally recognized just a century ago. It owes its existence, however, to the earliest British settlement in this west Caribbean region. During colonial times, this zone was sparsely populated by the Spaniards, who attempted to colonize a region historically dominated by the Mayas. The absence of valuable mineral deposits to generate quick returns, along with its relative isolation due to the difficult navigation across the world's second longest coral reef and dense tracts of forest, contributed to the marginal character of this region.

In the territory which is now Mexican Quintana Roo, the Spaniards had only one settlement, the village of Salamanca de Bacalar, founded in 1545 by Melchor Pacheco (Hoy 1983: 20). This isolated community, located next to the lagoon bearing the same name, was attacked and destroyed several times until the eighteenth century, when Marshall Antonio de Figeroa y Silva ordered the construction of the fort presently known as 'El Castillo de Bacalar' (Hoy 1983: 23).

The situation was different in what is now Belize. First, because the initial settlement came about by accident – a shipwreck which, in 1638, allowed the establishment of a community on the bank of the Belize River. Second, the discovery of a secure and rich timber area by a group of British pirates favoured the establishment of a timber enclave. This gave way to the consolidation of this region during the seventeenth century as a logging settlement. These settlers dedicated themselves mainly to the extraction of dyewood instead of piracy which, since the signing of the Treaty of Madrid in 1667, England had decided to cease supporting (Paz Salinas 1979: 21).

72

During the eighteenth century, the Spaniards and British signed agreements and, notwithstanding, engaged in numerous confrontations in this remote region. These conflicts apparently ended with the battle of Saint George's Cay in 1798, when the Spanish empire was in decline (Bardini 1978: 24). This border region was, from the end of the seventeenth century to the nineteenth century, the focus of a new type of exploitation which had replaced dyewood extraction because of the arrival of new synthetic dyes. The extractive economy of Belize was then geared to exploiting mahogany, coinciding with the boom of railway and ship construction in England.

Notwithstanding the importance of lumber extraction, the population in British Honduras was stagnant and unconsolidated, perhaps because the settlers of the interior, isolated from the coastal settlements, lived only on logging activities.

The event that was to change the situation of this region radically was the Caste War, a conflict between the Mayan population and the Yucatecos which broke out in 1847. This war was initially interpreted as an internal revolt, but today, because of its short- and long-term consequences, it is being reconsidered from a new perspective. It is now considered that, in addition to this conflict, there existed a complex

Table 6.1 Mahogany production in British Honduras, 1788–1846

Year	Volume of production (feet)
1788	5,271,275
1789	6,054,215
1797	2,081,000
1798	1,347,000
1799	3,550,000
1800	3,102,000
1801	3,061,000
1802	2,250,000
1803	4,500,000
1805	6,481,000
1824	5,550,000
1837	8,500,000
1840	4,500,000
1845	9,919,507
1846	13,719,075

Source: Bolland 1988; Handbook of British Honduras, 1890, 1898.

regional strategy devised by Britain on the Caribbean coast of Central America, extending from Nicaragua to Cape Catoche, the consequences of which are still visible today. From the Mosquito Coast in Nicaragua, where a king was imposed, to the Republic of the Macehuales on the eastern coast of Yucatán, Britain and the United States competed for influence through arms smuggling and conflict of interests. A solution was finally reached in favour of the United States with the Clayton–Bulwer Treaty of 1850, which intentionally did not include the issue of Belize.

The 1847 Caste War enabled Britain to consolidate its colony in Belize. Its creation was, paradoxically, supported by the Mexican peninsular population. Two basic premises emerged from the official creation of British Honduras in 1862. First, it secured an important population nucleus, large enough to settle and consolidate the most conflictive area of the colony, the northern border region with Mexico. This was made possible because Yucatecan emigrants, fleeing the war by crossing to the southern part of the Rio Hondo, received British protection. Likewise, on the northern side, the Mayas received ammunition from the British and, by the same token, an implicit recognition of their rights as a community.

Second, economic diversification was achieved and, most significantly, an important quota on food supplies was generated by these immigrants. In the process, the districts of Corozal and Orange Walk were created, the most important districts to this date. In 1861, a census carried out in Belize reported a total of 25,652 inhabitants of whom 52 per cent (13,542) were Yucatecan refugees (Ayuso 1986). This growing border zone was later to be known as communities of Mexican origin, namely Corozal, Blacklanding, Orange Walk, San Esteban and others. An 1865 census of this region reported 8,500 acres of land under sugar cane, bean, corn and cotton cultivation (Dachary et al. 1989a).

This situation encouraged the British Crown to impose legislation which restricted the actions of the immigrants. Two years later, a law was passed which deprived the Mayas and other exiles from the war of the right freely to occupy land by imposing a land tax to the Crown (Dachary et al. 1989b). The half-century of existence of the Macehuales Republic, and the utopia of the Mayas in the forests of the present-day Quintana Roo, permitted the consolidation of the colony of Belize, thus providing the basis for a border treaty.

During this period, an important group of refugees from the Caste War settled the three islands off the coast of Yucatán: Cozumel, Mujeres and Holbox. Through this action, they protected the maritime border by establishing commercial relationships in fishing with Cuba, Jamaica, the USA, England and Honduras.

In the 1880s a new situation arose in Mexico. The consolidation of capitalism in the country and the profitable sisal fibre production in Yucatán made necessary the stabilization of the south-eastern border of Mexico. This fact strengthened the government's objective of extending its control over the vast riches in the hands of the insurgents. Consequently, in 1893 the Spencer–Mariscal Treaty was signed, defining the present boundary between Mexico and Belize.

HALF A CENTURY OF BORDER RELATIONSHIPS

The signing of the Spencer–Mariscal Treaty opened the way to a settlement of the war. The war officially ended at the turn of the century since, in the treaty, Britain agreed to stop supplying arms to the rebels. However, initiating studies on its border with Belize, Mexico faced difficulties due to the paucity of information on it, a fact compounded by more than half a century of war. The Bacalar Chico Canal, the principal divide between the maritime zone and Chetumal Bay, is not a geographical accident, but rather an artificial waterway which Mexican fishermen at Ambergris had dredged, converting this region into an island (Dachary et al. 1989a) (see Figure 6.1). As a consequence, Mexico lost its exit to the sea and its ports in the southern zone. To this day, this situation persists without a solution. As a result, all imports to supply the south, and exports from it, are forced to transit through Belize.

With this situation, Mexico lost control over its resources for many decades. During this period, important lumber companies exploited the forest and transported its product from the Bacalar Lagoon through the Rio Hondo to Chetumal Bay and finally to Belize, from where it was shipped abroad.

By the end of the nineteenth century, an enclave economy had been consolidated on either side of the boundary (Dachary, 1986). During this period, the Federal Territory of Quintana Roo was created, administered from the centre of the country in spite of the ambitions of the local Yucateco oligarchy.

The mahogany industry gave way to the extraction of chicle gum. Powerful transnational companies, the majority from the USA, controlled these products, which continued to be exported via Rio Hondo–Belize. Through this means, the border region emerged as a major export centre for forest extraction products.

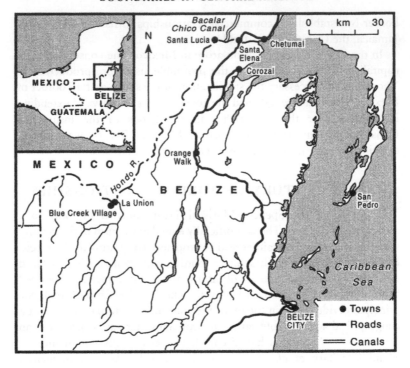

Figure 6.1 Mexico–Belize border region

The border started to be resettled. Once the war was over, the populations of the remote areas of the river were relocated next to the chicle or timber areas with their ports on the banks of the river. Located adjacent to these were the first customs offices for the control of the permanent transit of goods and people.

The border area developed in a very particular way during the first thirty years as a Federal Territory. The economy of this border region was practically led by and directed from Belize, headquarters of the major trading posts. Contracts were signed in Belize, and payments made there were in Belizean currency. The legal procedures associated with labour and production-related trials were also held in Belize. Meanwhile, at the border, the population lived a daily life ignoring the true intricacies of the boundary.

Extractive activities continued to be the mainstay of the region. They gave much dynamism to this area which was marginally located with respect to the more populated and economically articulated zones of

Table 6.2 Trade balance of customs in Chetumal, July 1898–May 1899

Month	Import ($)	Export ($)
July–Dec 1898	6,000	108,797
January 1899	349	2,006
February 1899	4,938	4,360
March 1899	3,665	12,838
April 1899	1,842	10,107
May 1899	2,239	24,240

Table 6.3 Maritime transit in Chetumal, 1905

Type	Entries			Exits		
	No. of ships	Tonnage	Crew	No. of ships	Tonnage	Crew
Steamer	63	1,610	346	64	1,622	348
Sailboat	285	5,643	180	279	5,687	837
National	57	875	1,178	53	894	183
Foreign	291	6,378	998	290	6,415	1,002

Source: Secretaria de Fomento, 1906.

Mexico. Commerce with Belize was extensive, principally in chicle gum and mahogany which were the main products of this border region as can be seen in Table 6.5.

During the world economic crisis of the 1930s, the border region suffered a sudden reduction in production, with the consequent reduction in population, loss of territory and isolation. This situation was solved only by its permanent linkage with Belize. In 1934, a commission arrived in Chetumal to study the effects of the crisis in this neglected border region, which on both sides was agonizing. The Irigoyen Commission, named after its coordinator, recommended the following: the establishment of free trade zones; free traffic for foreign ships; reduction of the transit fees; subsidies for air transport. Finally, it recommended a set of measures to revitalize the economy of this forlorn section of the southern border.

A year later, under the Cárdenas presidency, the economy of the territory was revitalized. The chicle boom gave it the necessary momentum, bolstered by increased market demand during the Second

Table 6.4 Location of custom offices

Place	Distance
Rio Hondo region	
Santa Elena	12 km from Chetumal
Chac	8 km from Santa Elena
Santa Lucia	3 km from Chac
Ramonal	11 km from Santa Lucía
Menguel	17 km from Ramonal
Botes	34 km from Menguel
Blue Creek	38 km from Botes
Rio Hondo's total distance	123 km from Chetumal
Chetumal Bay region	
Calderitas	12 km from Chetumal
Coastal Caribbean region	
Xcalk	80 km from Chetumal (by boat)

Source: 'El problema económico en Quintana Roo', 1934.

Table 6.5 Export to Belize from the maritime and border customs in Chetumal, 1918–24

Year	Mahogany (cu. m)	Chicle (kg)	Fruit (kg)
1918	742,881	266,619	14,600
1919	2,135,459	352,844	10,355
1920	5,289,953	442,154	–
1921	2,895,856	326,797	21,283
1922	7,125,789	162,195	–
1923	–	328,636	15,000
1924	7,827,181	256,907	35,075

Source: Informe de la Comisíon Aguirre, 1925.

World War. During this period, the three zones into which the border is divided were finally conformed, namely: the Caribbean area, which includes Xcalac, Mexico and San Pedro, Belize, with an economy based on copra and fishing (Dachary et al. 1988); a second zone located on the Bay with an economy based on commerce and fishing, which includes the two major communities of the zone, Chetumal, Mexico and Corozal, Belize; and, finally, the Rio Hondo zone, based on the forest enclave, consisting of numerous communities on both banks extending from the source to the mouth of the river.

THE CONTEMPORARY PERIOD

By the end of the 1950s, the forest enclave in Mexico started to decline. In Belize, a similar situation had already begun. Corozal succeeded in replacing it with sugar cane which had started to be processed at a mill located in the region, at La Libertad.

The Belizean border zone experienced extensive changes during the 1960s as a consequence of the arrival of the English transnational company Tate and Lyle which, faced with the depletion of cane plantations in Jamaica, acquired the mill at La Libertad on the border zone and encouraged the expansion of this agricultural zone (Alvarez 1987). During this period, the first groups of Mennonites arrived at Blue Creek, located at the far western zone of the border between the Rio Hondo and Blue Creek, stimulating the resettlement of this former forest region.

In Mexico, the movement towards the sea was encouraged. A reformulation for the use of its coast and its borders guided the first resettlements in this region. However, it failed due to the lack of support and effective planning. But the decades of the 1960s and 1970s produced a series of events both national and international which encouraged the Mexican Government to focus its attention on this neglected south-eastern zone of the country.

On the international scene, two events caused upheaval in the Caribbean region: first, the Cuban Revolution and, second, the initiation of a process of struggle for independence, especially of British colonies, a process which did not exclude Belize. The insurrection movements of the 1970s deepened the problems, principally at the Belize–Guatemala border zone. At this time, the latter began a campaign of reclamation to impede the independence of Belize. Guatemala even reached the extreme of threatening Belize with the use of force to take over its territory.

At the national level in Mexico, two important events changed the geopolitical vision of the region, first, the discovery of vast deposits of petroleum in the south-east, particularly in Campeche, and, second, the return of Mexico to the Caribbean region with the gigantic tourist project, Cancun, filling the void left in the region by revolutionary Cuba. For Mexico, both events implied a reformulation of their southern border, since accelerated demographic change was expected in the region. In the case of the border with Belize, a planned process of settlement began in the 1970s, which succeeded in resettling the whole area within a decade. Fourteen new population centres, with a total of twenty communities, settled in the bordering muncipality of Othon P. Blanco. This permitted the settlement of more than 10,000 persons with the objective of integrating them in the sugar cane agroindustrial project. In 1981 Belize obtained its independence and with this event a long era of colonialism in the region came to an end.

THE PRESENT-DAY SITUATION

The commercial relationship which has existed between Mexico and Belize since the beginning of the century has become stronger as a consequence of the opening of a free zone in 1976. This fuelled Chetumal's economy until 1990 when, because of the GATT agreements, imports by Mexico sky-rocketed, causing the free zones to enter a period of crisis.

Since the devaluation of the Mexican peso in 1982, the flow of border trade for regional consumption has been inverted. Chetumal has become the main supplier of food and manufactured goods for the northern and central populations of Belize. At present, this trade represents, for the commercial capital of Quintana Roo, an important injection and an alternative to the negative effects of GATT.

This informal border trade, as known in the border zone, has the following characteristics:[1]

1   The flow of visitors is large, with an annual daily average of 2,000 persons crossing the border, which is high when we take into account the fact that Belize has only 184,340 inhabitants (Central Statistical Office 1990).

2   The average expenditure is more than 150 American dollars per person, and purchases are for reselling, for consumption and other purposes, medical expenditures featuring as a high priority.

3   Petrol consumption is very high, an average of 300 litres per person, between 350 and 400 per cent more than the normal average of Chetumal City. The principal reason for this is that petrol in Mexico is about one-third of the price of petrol in Belize.

4   The majority of the visitors come from the most populated districts of northern Belize – Corozal and Orange Walk – and their average return is within ten days.

In addition to this informal trade, a significant illegal trade exists, mainly in household items, petrol and medicine, which enters through Rio Hondo. Mexican exports to Belize are basically construction materials: cement, iron and sanitary equipment. These Belizean imports have been increasing since 1985, the year in which the first records were registered at the Custom Office in Progreso, Yucatán. At present, the border region is experiencing a tourist boom developed by Belize. In San Pedro, the Belizean border zone on the Caribbean coast, there are 500 rooms, and in the Bay zone, 100, competing with approximately 1,000 rooms in Chetumal and Bacalar (Central Statistical Office 1990). Common zones of exploitation are identified in the agricultural sector, on both sides of the border. In the Rio Hondo zone there is an extensive sugar region, on the Belizean side, centred in the factory of Tower Hill situated in the Orange Walk District. On the Mexican side, the sugar region has as its centre the mill at Alvaro Obregón. Towards the west is the livestock region, found on both sides of the river, the Belizean side being the most developed, having as its centre the Mennonite community.

Close to Peten, Guatemala, are the forest zones, exploited on both sides of the border with a concentration on precious woods such as mahogany and cedar. Along the coast, fishing cooperatives can be found on both sides, though they are in a process of transition-reduction caused by the tourist boom.

THE EMERGENCE OF NEW CONFLICTS

Two important conflicts are altering the border situation. These are drug trafficking and illegal immigrants.

During the 1980s, the districts of Corozal and Orange Walk registered increased activity in the production and commercialization of marijuana. At the same time this was experienced, though to a lesser extent, in the muncipality of Othon P. Blanco in Mexico. This generated permanent violence in the region and conflicts between the traffickers and the authorities. Since 1985, Mexico and Belize have been collaborating occasionally. In 1990, an agreement was signed to cooperate, in a coordinated fashion, in the war against drug trafficking. The fumigation of drug plantations at the border region has been a constant source of conflict since it is claimed that the sovereignty of the countries fumigated is not being respected. At present, the conflict arises more from trafficking than from the production of drugs. The other conflict zone produced by drug trafficking is the maritime border along the Caribbean coast. This zone is located strategically as a major route for the passage of drugs coming from South America.

In general, the southern border has two types of illegal immigrants: the bordering zone between Mexico and Guatemala is crossed mainly by Central Americans, while the Mexican–Belize zone registers a low rate of illegal Central Americans but a high number of illegal Asians and Africans. These come mainly from the British colonies which have easy entrance into Belize and whose nationals coincide with the different races found in that country.

REGIONAL PERSPECTIVES

Over the last two decades, many important changes have taken place in the region, and have had far-reaching consequences. The northern Belize border region with Mexico is the most populated and fertile zone of Belize. The diversification towards tourism occurring in this region, though only on its Caribbean coast, makes this region a development priority for Belize. The dynamic border relations brought about by tourism and the increased commercial relationships have generated a major development of the means of communication: new roads, air routes and so on. An agreement on open skies now permits direct flights between Belize and Cancun and Belize and Chetumal. In addition there are bus lines connecting the two major cities, Belize and Chetumal, with more than ten buses per day. In addition to these tangible relationships, there are other potentials that can be explored. A common export-processing zone on the banks of the Rio Hondo is being developed with Taiwanese capital, and many joint venture projects between Mexico and Belize are feasible.

A potential petroleum project is pending following the Interparliamentary Mexico–Belize Meeting (September 10, 1991). The Belizean delegation at this meeting requested the presence of PEMEX (Mexican Petroleum) to drill the border zone, adjacent to Peten, where important deposits of oil have been detected.

These and other actions brought about by the economic boom and by demographic growth have radically altered the potential of this neglected border. Important changes resulting in new relationships between Mexico and Belize are expected during the 1990s. These new border relationships will constitute important elements for the Central America integration process. Though Belize has not been included in this process, there are high probabilities that in the near future it will be. An accelerated change in the demographic and ethnic structure suggests for Belize a stronger integration with the region, of which today the border zone is at the forefront.

Thus a border region neglected for a century now faces a new direction amidst the development process of the complex Central American region.

NOTES

1   Data collected for the Informal Border Trade Project, through surveys and census carried out between April and July 1990.

BIBLIOGRAPHY

Alvarez, P. (1987) *Belice, la crisis del neocolonialismo y las relaciones con Mexico 1978–1986*, Mexico City: Cide.
Ayuso, M. (1986) *The Role of the Maya – Mestizo in the Development of Belize, 200 BC to 1984*, Belize City: Mimeo.
Bardini, R. (1978) (ed.) *Belice historia de una Nacion en movimento*, Tegucigalpa, Honduras: Universitaria.
Bolland, N. (1988) (ed.) *Colonialism and Resistance in Belize*, Belize City: Cubola Production.
Central Statistical Office of Belize (1990) *Proyecciones preliminares*, Belize City: Government Publication.
Dachary, A. C. (1986) *Las etapas del desarrollo economico, le monde diplomatique, En Castellano*, Mexico City: Enero.
Dachary A.C., Dachary, A.B. and Maris, S. (1988) *Estudios socioeconomicos preliminares de Quintana Roo sector Pesquero*, CIQRO (Centro de Investigaciones de Quintana Roo), Cancun: Tomo V.
—— (1989a) *Nuevas fronteras repoblamentio y formacion del capitalisimo en la Costa Oriental de Yucatan*, unpublished CIQRO paper from VIII Seminario la formacion del capitalismo en Mexico.
—— (1989b) *El Caribe Mexicano una introduccion a su historia*, Mexico City: Fondo de Publicacionese de Quintana Roo.
Hoy, C. (1983) *Breve historia de Quintana Roo* 2nd Edition, Mexico City: Chetumal.
Paz Sallinas, M.E. (1979) (ed.) *Belize, el despertar de una nacion*, Siglo XXI, Mexico City: Tegucigalpa.

7

THE INTEROCEANIC CANAL AND BOUNDARIES IN CENTRAL AMERICA

The case of the San Juan River

Pascal O. Girot[1]

INTRODUCTION

There are few regions of the world where physical geography has influenced so markedly the political destiny of nations and states as in Central America. Its isthmian configuration bestowed upon it a potential for interoceanic communication which in turn has fostered for centuries the ambition of empires and enterprises alike in their struggle to control a passageway across this 'thin waist of the Americas'. This struggle has expressed itself in long cycles of rivalry between the hegemonic powers for the control of an interoceanic route across the Central American isthmus.

On the other hand, the comparative geographic advantage of a particular interoceanic route versus another has pushed Central American states to offer their territories for the grand enterprise. This rivalry between neighbouring states has often expressed itself in conflicts over territories and water resources potentially to be used for an interoceanic canal. Such is the case of the San Juan River between Nicaragua and Costa Rica. In the case of Panama, the limited exercise of its sovereignty and the strict control over water resources feeding the canal by the United States illustrate the territorial implications of the control and administration of an interoceanic waterway.

These examples suggest that the canal imperative has had a particular influence on the political make-up of Central American states. One could argue that the 'Balkanization' of the Central American isthmus has been fuelled in part by the transisthmian prerogative. The territorial division of Central America into sovereign states, each with its relative potential for an interoceanic route, has constituted one of the

84

main obstacles to the economic and political integration of the region. Thus one could finally suggest that the transisthmian prerogative has been dialectically opposed to the process of regional integration of the Central American states.

To illustrate these points, I intend to examine critically the relationship between canal routes and Central American boundaries, focusing in particular on the case of the San Juan River between Nicaragua and Costa Rica. Following a broad comparison of the major canal routes across Central America, the particular case of the San Juan River–Lake Nicaragua route will be detailed. The historical sequence of canal studies and proposals lends support to the persistence of the transisthmian imperative, particularly during the second half of the nineteenth century. The relationship between the canal route proper and the boundary demarcation separating Nicaragua and Costa Rica will also be assessed. Finally, a reflection on the territorial and geopolitical implications of the canal prerogative for Central America will be provided.

CENTRAL AMERICA'S INTEROCEANIC PASSAGEWAY: A GEOPOLITICAL IMPERATIVE

Ever since Balboa stumbled upon the shores of the Pacific Ocean in 1513, the Central American isthmus has been the object of recurrent rivalries between hegemonic powers competing for control over a potential interoceanic route. This could partly be explained by the universal concern for speeding up communications across an ever-shrinking world to serve an ever-expanding global economy. As early as 1550, the Portuguese navigator Antonio Galvao identified the three major transisthmian routes: Tehuantepec, Nicaragua and Panama (see Figure 7.1).[2] The Spanish Crown soon capitalized on the enormous potential of the Central American isthmus. Tehuantepec and Panama were employed throughout the Colonial period as land bridges, linking the Spanish Main[3] to Peru, Chile and the Philippines.

Alexander von Humboldt, in his *Political Essay on New Spain* published in 1811, underlined the importance of the Central American isthmus for global communications. In his work, he compared the five major sites for an interoceanic passage, including Tehuantepec, Nicaragua, and three locations in Panama and New Granada (San Miguel, Caledonia Bay and the Atrato River-Uraba gulf).[4] Humboldt's work constituted a major watershed in the European scientific and political community. His appreciation of Central America's potential

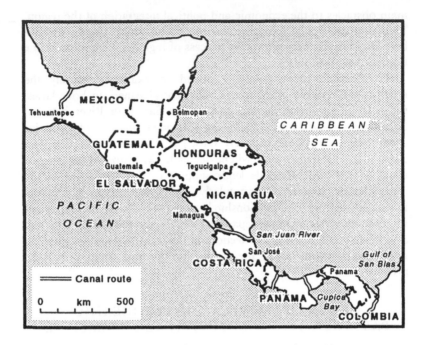

Figure 7.1 Transisthmian routes in Central America

for an interoceanic canal spawned heightened interest from European powers, and stimulated a series of surveys and projects during the nineteenth century. Inspired by Humboldt's ideas, Simon Bolivar, as President of Greater Colombia, commissioned in 1828 two European engineers, the Briton John A. Lloyd and the Swede Maurice Falmark, to survey the isthmus of Greater Colombia and to propose a feasible interoceanic route.[5]

This survey opened the way for a whole series of similar feasibility studies across the Central American isthmus, initially by Europeans and, during the second half of the nineteenth century, by Americans. The major routes surveyed at the time were the Lake Nicaragua–San Juan River and the three sites of the isthmus of Panama. In addition to their technical aspects, which evolved with the improvement of geodetic and cartographic techniques, these surveys provided firm grounds on which to base the geopolitical designs of the major world powers at the time. If one analyses the sequence of surveys and the governments by which they were commissioned, one can reconstruct the great episodes of imperial rivalry that marked the nineteenth century. First came the

86

French and the British, and after 1850 the canal surveys were accomplished almost exclusively by American engineers, at first private entrepreneurs, then government technicians.

Geopolitical interest in the Central American canal culminated during the latter part of the nineteenth century, specifically from 1879 onwards. The 1880s and 1890s brought about a 'scramble for the canal' concomitant with the famed 'scramble for Africa', with, as a major difference, a strong participation by the United States. As a rising world power, the United States competed with European powers, chiefly France, and to a lesser extent Britain, for concessions, surveys and political control of an interoceanic route across the Central American isthmus. Ferdinand de Lesseps, the French engineer responsible for the Suez canal, led the disastrous attempt to construct a sea-level ship canal across Panama in 1880.[6] With the French actively building the Panama Canal during the 1880s, the United States Government commissioned a number of surveys along the Nicaragua route, until then chiefly undertaken by private concerns.

As a corollary to the gathering of technical data, both the United States and France sought legal concessions and leases to guarantee control over the operation and defence of the future canal. As a result, a series of treaties and protocols were signed both between the world powers (Britain and the USA), and between these major powers and Central American states. The 1846 Mallarino–Bidlack Treaty between the United States and New Granada provided the former with the right to free transit across the isthmus of Panama. The Clayton–Bulwer Treaty of 1850 celebrated the 'gentlemen's agreement' between the United States and Britain in which each party agreed not to seek to gain exclusive control over any future Central American canal. A large number of treaties were negotiated during the second half of the nineteenth century. Of particular note was the Cañas–Jeréz Treaty of 1858, between Costa Rica and Nicaragua, providing the latter with total control over the waterway encompassing Lake Nicaragua and the San Juan River. In order to proceed with the initial excavation of the Panama Canal, the French secured a concession for the work with the Colombians through the Salgan–Wyse Contract. Following the failure of the French venture in 1889, this concession was prorogated in 1890 by the Roldán–Wyse Contract, thus maintaining the French presence in Panama. This fact was to be of paramount importance in the final decision in favour of the Panama option by the United States.[7]

The 1890s brought about a culmination in the geopolitical interest in an interoceanic canal across Central America. The fiasco of the French venture in Panama only bolstered the resolve of the United States Government to gain control over any future canal project. Two main obstacles remained. First, the concession over the Panama route was still in French hands, in the form of the Roldan–Wyse Contract. Second, the 1850 Clayton–Bulwer treaty between Britain and the United States still made the construction of a canal by the latter conditional upon consultation with the British. As an emerging world economic power, the United States had an ever-increasing interest in controlling trade routes across the Central American isthmus without being impaired by any European power. This tendency was compounded by the territorial consolidation of the United States as a continental power during the second half of the nineteenth century, controlling territories from Texas to California. Following her victory in the Spanish–American war of 1898, the United States consolidated her hegemonic position, gaining control over Cuba, Puerto Rico and the Phillipines.

Inspired by the thesis of Alfred T. Mahan, arguing for the supremacy of American naval power, the administration of Theodore Roosevelt took definitive steps towards consolidating its hegemonic control over the Central American isthmus. An array of bilateral treaties and legal provisions opened the way for the construction of 'an American Canal on American soil' (McCullough 1977: 257). These included the 1901 Hay–Pauncefote Treaty between Britain and the United States which replaced the binding Clayton–Bulwer Treaty, allowing for the exclusive political control of an interoceanic canal built and defended by the United States. The Spooner Law, passed by the US Congress and Senate in 1902, gave the Roosevelt presidency the necessary backing for the construction of an American canal in Central America. With the rejection of the 1903 Herrán–Hay Treaty by the Colombian Parliament, and the creation of Panama that same year, the United States mustered the necessary conditions for the construction of the Panama Canal under exceptionally advantageous conditions.[8]

The Hay–Bunau Varilla Treaty of 1903, between the United States and the newly independent Panama, profoundly modified the nature of the concession over the canal route previously fixed by the Herrán–Hay Treaty. First, the concession was granted in perpetuity, instead of for renewable 100-year periods. Furthermore, the Canal Zone under United States sovereignty was expanded from six to ten miles on either side of the proposed route, including unlimited access to land and waters for the functioning of the canal (initially fixed to a maximum of fifteen miles from the waterway) (Castillero 1954: 131). This aspect of the Hay–Bunau Varilla Treaty reinforces the notion that the territorial

88

control over land and water was an essential ingredient of the political cost of a canal concession under such hegemonic conditions. Despite further amendments and adjustments to the 1903 treaty by Panamanian lawmakers, the basic tenets allowing for the full exercise of sovereignty by the USA over the canal route and its water supply have remained.[9] The Carter–Torrijos Treaty of 1977 was aimed precisely at ending US territorial control over the Canal Zone and restoring it to Panamanian sovereignty by 1999. Figure 7.2 illustrates the territorial concessions linked to the Panama Canal Zone and the influence which hydropolitics has had on the delimitation of US jurisdiction over waters supplying the canal, constituting a *de facto* territorial enclave across the Central American isthmus.

The particular territorial configuration of the Panama Canal provides an interesting basis on which to compare the other major canal route in

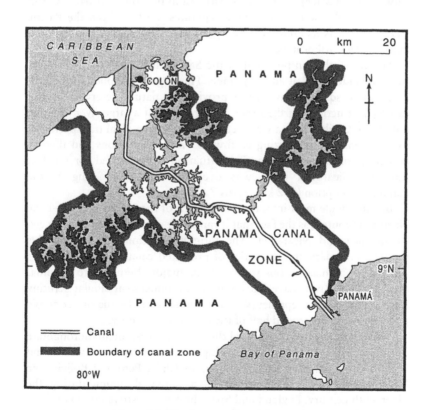

Figure 7.2 The Panama Canal Zone, 1990

Central America. The San Juan River–Lake Nicaragua route provides a compelling case study illustrating the relationship between the canal issue and boundaries in Central America.

THE SAN JUAN RIVER AND THE INTEROCEANIC CANAL ROUTES

As the most extensive watershed in Central America, the San Juan River–Lake Nicaragua drainage basin covers over 40,000 sq. km, two-thirds in Nicaraguan territory and a third in northern Costa Rica. It owes its particularity not only to the fact that it is a binational watershed, like many others in the region, but essentially to its unique transisthmian position. Draining the Nicaraguan volcanic depression (or Graben), Lake Nicaragua constitutes the largest freshwater body between Lake Michigan and Lake Titicaca in the Andes. A narrow strip of land merely twelve miles wide separates the lake from the Pacific Ocean, whereas its outlet, the San Juan River, flows over two hundred miles to the Caribbean Sea.

The transisthmian potential of the San Juan River–Lake Nicaragua route was identified at an early stage. The large water body of the lake (over 8,000 sq. km) was seen as providing a natural reservoir for any canal construction. Furthermore, the presence of a large river draining out to the Caribbean Sea was seen as providing an ideal transportation axis, potentially minimizing earthworks and excavations and thereby reducing construction costs. This situation compensated for the fact that the Nicaragua route was over three times as long as the Panamanian option (292 km versus 72 km).

But the single most important drawback of the San Juan River–Lake Nicaragua route was the fact that it ran along a political border between Costa Rica and Nicaragua. The intimate relationship between this political boundary and the series of projected canal routes constitutes the major emphasis of this study. As the major alternative to Panama, the San Juan River–Lake Nicaragua route underwent similar scrutiny from engineers, entrepreneurs and politicians of the major economic powers during the second half of the nineteenth century.

While Panama had constituted the major transisthmian thoroughfare during the colonial period, channelling to Spain much of the wealth and bullion extracted from its possessions in Chile, Peru and Bolivia, the Nicaraguan route truly emerged as a potential option during the nineteenth century. England and Spain had earlier striven to control the route and, by the mid-nineteenth century, the San Juan route became the focus of world attention.[10] The California Gold Rush, starting in 1848, brought about a major transformation in the United States' perception of the Central American isthmus, in the absence of any modern overland transportation across North America. Both the Panamanian and the Nicaraguan routes became major routes of passage between the east and west coasts of the United States.

The completion of the interoceanic railroad across Panama in 1855 provided further incentives for consolidating control over a potential ship canal route. By the last quarter of the nineteenth century, geopolitical interest in a canal route across the Central American isthmus had peaked. The 'scramble for the canal' took the form of numerous technical surveys, concessions and contracts in which the major economic and political powers of the moment participated. The British and the French in the first instance, and the United States by the 1870s, were major players in the rivalry over the control of a future transisthmian route. The San Juan River–Lake Nicaragua route constituted a major staging ground for this rivalry. Between 1850 and 1901, six major canal proposals were formulated; four of them based on extensive fieldwork and surveys of the canal transect. Figure 7.3 illustrates the proliferation of canal projects and routes across the border region.

Among the earlier proposals was that by Orville Childs, contracted in 1850 by Cornelius Vanderbildt, who at the time ran the Accessory Transit Company taking passengers across the Nicaraguan isthmus (Childs 1851; Williams 1971). The proposal was sent to British bankers but to no avail. In 1858, a Frenchman by the name of Félix Belly also formulated a proposal for a ship canal across the San Juan River–Lake Nicaragua and Lake Managua route, playing on concessions from both Costa Rica and Nicaragua.

The aftermath of the Filibuster War of 1856–7 opposing American mercenaries and Central American troops provided the basis for major agreements between Costa Rica and Nicaragua. The 1858 Cañas–Jeréz Treaty fixed the present-day boundary between Costa Rica and Nicaragua. As will be detailed later, the treaty established complete Nicaraguan sovereignty over the course of the San Juan River.

The 1860s were marked by a relative lull in the pace of canal surveys. In 1863, an offspring of Félix Belly's plans, the American Atlantic and Pacific International Ship Canal Company, headed by Edward Loos and grouping French and British interests, proposed a canal across Nicaragua, spanning both Lake Nicaragua and Lake Managua to reach Tamarindo on the northern Pacific coast of the country. The project

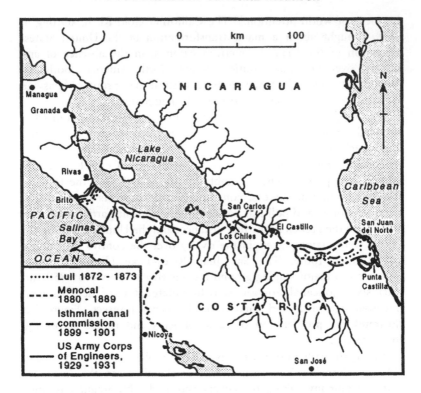

Figure 7.3 Principal canal routes across the Nicaraguan Isthmus

provided for no less than fourteen locks on the Pacific side of the Canal, involving the removal of over 6.3 million cubic yards of earth for the excavation of a 32-foot deep trench between Lake Managua and the Pacific coast (Belly 1867; Loos 1863). The entire proposal derived from no direct surveying or gathering of technical data, and was plagued by gross miscalculations of the topography and hydrology of the region. The failure to attract even minimal financing led to the dismissal of the project as a whole. This was to mark the end of European, chiefly British and French, attempts at building a canal across Nicaragua.

By the 1870s, the United States Government began to show marked interest in the potential for an interoceanic canal across Central America. A surveying party headed by Edward P. Lull in 1872–3 was commissioned by the US Navy Department to provide the first serious scientific compilation of topographic and hydrological data on the San Juan River. The proposal was for a $65.7 million canal across the Rivas isthmus and along the San Juan River (U.S. Congress, Senate, Navy Department 1874: 98). Departing from previous studies, it paid close attention to the problems inherent in overcoming the presence of rapids along the middle course of the San Juan River. Similar attention was paid to the problem of excavating an artificial canal across the Rivas isthmus. Several dams were planned below the mouth of the San Carlos River, a major tributary of the San Juan River, to elevate waters and thereby eliminate the problem posed by rapids. However, although the Lull survey provided adequate topographic information, scant attention was given to the regime and siltation inherent in tropical rivers such as the San Juan, a problem which was to be paramount in later canal surveys.

While the Lull proposal failed to secure financing for such a prohibitively costly venture, one of the engineers who participated in the surveys, Aniceto Menocal, spearheaded further surveys of the San Juan route in the 1880s. His 1885 proposal substantially improved on the early Lull project (Menocal 1890). Reducing the number of locks and dams, he managed substantially to lower the final cost of construction. Menocal also sought to solve the problem of siltation in the lower course of the San Juan by circumventing it completely, running the canal northward toward San Juan del Norte along the Deseado River. Instead of several dams, Menocal advocated a single large dam downriver from the mouth of the San Carlos River, the largest tributary of the San Juan, which would also feed the lower portion of the canal. This giant dam was to eliminate the problem of rapids, overcome seasonal surges in river discharge, and limit the siltation of the canal (U.S. Congress, Senate, Navy Department 1886: 27). However, such a dam would have inundated much of the watershed, including large portions of Costa Rican territory, a fact which fuelled serious tensions between the neighbouring countries. The litigation culminated in 1888 with the boundary arbitration by US President Cleveland, which reiterated the terms set by the Cañas–Jeréz Treaty of 1858, giving Nicaragua full sovereignty over the waters of the San Juan River.

The Menocal proposal crystallized in 1889 when the Maritime Canal Company of Nicaragua was created, opening the way for the first and last attempt at canal construction along the San Juan route (Maritime Canal Company Of Nicaragua 1889 and 1890). Coinciding with the fracas of the French canal venture in Panama, the attempt at constructing the canal based on Menocal's survey lasted five years, and suffered a similar fate, though far less catastrophic in financial and human terms. The 1893 financial crisis in the USA paralysed operations, leaving several miles of railroad construction and a mile-long portion of the canal excavated to be inexorably conquered by the jungle. The rusty remains of some of the dredges left behind can still be seen in Greytown harbour to this day.

The last decade of the nineteenth century saw the culmination of the 'scramble for the canal', with an unprecedented number of surveys and commissions sent to Nicaragua and Panama to ascertain the feasibility of an interoceanic canal in engineering as well as economic terms. The U.S. Nicaragua Canal Board submitted in 1896 a report to the United States Government pointing out some of the weaknesses of the Menocal proposal (U.S. Nicaragua Canal Board 1896). The lack of systematic hydrological and geological data did not substantiate the location of the large dam proposed by Menocal. The board finally recommended more detailed data gathering, and close scrutiny of the eastern terminal of the proposed canal.

Thus was created the Nicaragua Canal Commission (NCC), led by John G. Walker, which completed the most thorough hydrological survey of the San Juan River ever carried out, including detailed measurements of river discharges, rainfall and geological formations. A lengthy report, submitted in 1899, concluded favourably in terms of the feasibility of canal, for a total cost of over $138 million (U.S. Nicaragua Canal Commission 1899: 46). The features of the canal design proposed by the NCC differed substantially from Menocal's earlier proposal, eliminating most of the dams and emphasizing much more extensive dredging and excavation of the entire route.

Finally, the Isthmian Canal Commission (ICC) was created (1899–1901) by the US Government to examine, compare and synthesize all the options in Central America for an interoceanic canal. The report submitted by the ICC constitutes the culmination of a century of explorations, surveys and feasibility studies for a canal in Panama and in Nicaragua especially. The ICC report, basing its arguments on the NCC report, finally favoured the Nicaragua route over the Panamanian option. The clinching factor was the total cost of both projects. Since French interests still controlled the Panamanian concession, offering to sell the installations abandoned over ten years earlier for the colossal sum of $109 million, this brought the total cost of the Panamanian option to over $250 million. This did not compare with the estimated cost of $189.9 million for the Nicaragua route.[11] Thus, despite the intrinsic advantages of the Panamamian route (shorter length, existing port and railroad facilities, and lower summit elevation), Nicaragua was seen to be 'the most practicable and feasible route for an isthmian canal to be under the control, management and ownership of the United States' (U.S. Congress and Senate 1904: 175).

The events of 1903 were destined dramatically to change these circumstances. The political secession of Panama from Colombia, and a last minute rebate of the canal concession by the French, turned the tables in favour of the shorter option. Thus the United States secured an exceptionally advantageous territorial concession, under no less exceptional political terms. The San Juan route, touted only two years earlier as the option for an interoceanic canal, fell quickly into oblivion. The town of San Juan del Norte, which had mushroomed during the 1890s, soon waned into a sleepy and desolate outpost of the San Juan Delta.[12]

THE DIALECTICS BETWEEN BOUNDARY AND CANAL DELINEATIONS

Perhaps the single most important feature which distinguished the San Juan route from the Panamanian option, aside from the technical aspects of length, summit elevation and hydrology, was the fact that the proposed canal project ran along the international boundary separating the states of Nicaragua and Costa Rica. This fact offers a particularly interesting context in which to study the interrelation between territory, water resources and boundary definitions in the Central American isthmus. The cardinal influence of the canal issue in the fixing of the political boundary between Nicaragua and Costa Rica constitutes one of the most interesting examples of hydropolitics in a region marked by strong hegemonic control by external powers.

The mouth of the Desaguadero

The historical origin of the San Juan River as an administrative boundary dates back to 1541 when the Desaguadero (the river draining Lake Nicaragua) was mentioned as the division between the colonial provinces of Cartago and Nicaragua, both part of the Captaincy General of Guatemala. By the end of the sixteenth century, the mouth of the Desaguadero was defined as forming the northern limit of the Province of Costa Rica. Costa Rica's *Ley Fundamental* of 1825 mentions the San Juan River as the northern limit of its territory.[13] This definition remained as an ambiguous reference for future boundary demarcations, especially considering the constant geomorphological evolution of the San Juan River delta over time, as is well illustrated by Sandner (1987).

Notwithstanding the fact that the San Juan River was the setting of major military conflicts between the Spanish Crown and European pirates and troops during the colonial era (including the 1780 incursion by the British Navy in which Captain Horatio Nelson participated), the issue of the fluvial boundary between Nicaragua and Costa Rica truly emerged during the nineteenth century. After independence from Spain, the region initially formed part of the Central American Federation, until this collapsed in 1838. The 1840s marked the beginning of boundary disputes between Nicaragua and Costa Rica. In addition to the San Juan River Basin, the newly independent states shared a common feature: an extremely poorly integrated territory, particularly in the total absence of ties linking the more populated Pacific highlands to the Caribbean coast. This fact was compounded by the enduring British presence in, and control of, the Mosquito Coast to the north of the San Juan Delta (Dozier 1985; Peralta 1898).

The emerging economies of Costa Rica (whose principal export was the flourishing coffee trade) and Nicaragua led their governments to seek access to Caribbean ports, in order to avoid having to depend on the British merchant ships which circled Cape Horn (Gonzalez 1976: 57). One of the first boundary negotiations between Nicaragua and Costa Rica, in 1846, involved the latter's use of San Juan del Norte as a port for its coffee exports. Aside from the high tariff required by Nicaragua for Costa Rican use of the San Juan, the outcome of the negotiation linked any settlement to the precise delimitation of the boundary separating the two countries. Despite arbitration by Guatemala and Honduras, the treaty was not ratified by the Costa Rican parliament because of the excessive duties required on exports and imports by Nicaragua. An additional factor which influenced negotiation between Costa Rica and Nicaragua at the time was the annexation of the *partido* of Nicoya several years earlier, in which this large Pacific coast province voted to form part of Costa Rican territory, an act repudiated ever since by Nicaragua (Murillo 1986: 47).

During the 1840s, Britain sought to gain advantage from its unique position as the world's major economic and military power, and expand its territorial control from the Mosquito Coast southwards to the San Juan River Delta, vying for control over any future interoceanic canal across Central America. In 1848, the British Navy took San Juan del Norte at the mouth of the San Juan. This event coincided with a heightened interest in the potential canal route, and brought about a flurry of diplomatic activity from all parties. Costa Rica sought to settle its differences with Nicaragua by ceding two to three leagues of territory on the southern side of Lake Nicaragua and the San Juan River to facilitate Nicaraguan control over a potential canal route. In 1848, the Costa Rican Government commissioned Felipe Molina to lead negotiations with Nicaraguan government officials. Molina's mission was to seek a series of reciprocal trade agreements which could secure free transit on the San Juan and free use of the San Juan del Norte port facilities for Costa Rica, while reiterating total Costa Rican sovereignty over the Nicoya Peninsula (Molina 1851). As a result of these negotiations, Costa Rica secured full sovereignty over Nicoya and the southern tributaries feeding into the San Juan River, especially the Sarapiquí River which had been an object of contention. Costa Rica, however, renounced all rights over a future canal, and still had to pay high tariffs for the use of the port of San Juan del Norte.

Concomitant with the ongoing negotiations between Nicaragua and Costa Rica, Britain and the United States were vying for control of a future interoceanic canal. Seeking to stall British territorial claims on and beyond the Mosquito Coast, the Nicaraguan government eagerly sought the support of the United States. On the other hand, Costa Rica, playing on the rivalry between the two powers, strove for an alliance with Britain so as to obtain guarantees on the San Juan. This game of diplomatic alliances, in which imperial rivalries mirrored the local rivalry between Costa Rica and Nicaragua, came to a head in the late-1840s when Britain and the United States came as close as ever to all-out war, as did Nicaragua and Costa Rica (Sibaja 1974; Murillo 1986). Each country offered concessions for the building of the canal to the rival powers, thus further fuelling tensions between them.

By the end of the 1840s, US interest in the San Juan route had materialized in Cornelius Vanderbildt's Accessory Transit Co., which carried passengers between the east and west coasts of the USA across the San Juan route. Between the start of the California Gold Rush in 1848 and the opening of the continental railroad across the USA in 1869, over 150,000 passengers used the San Juan route to cross the Central American isthmus (Folkman 1972: 236). One of the largest industrial fortunes of North America was partly built on this transisthmian transport, when Cornelius Vanderbildt controlled the Accessory Transit Route across Nicaragua, combining steamboats and carriages to take passengers between the two coasts of the United States. The rivalry between Britain and the USA was finally settled in a diplomatic 'gentlemen's agreement' in the form of the previously mentioned Clayton–Bulwer Treaty, in which both parties renounced exclusive rights over any future canal venture.

The 1850s ushered in an even more tumultuous period in the region. The invasion and take-over of Nicaragua by US filibuster troops led by William Walker in 1855 led to all-out war, in which Central American troops united to defeat the invaders. Paradoxically, the Filibuster War brought Costa Rica and Nicaragua closer together than ever, united against a common enemy. Despite residual tensions created by the control of the San Juan River by Costa Rican troops as an outcome of the war, Nicaragua and Costa Rica reached a lasting settlement in the form of the Cañas–Jeréz Treaty.

THE CAÑAS–JERÉZ TREATY AND THE BOUNDARY DELIMITATION

The Cañas–Jeréz Treaty of 1858 remains to this day the essential basis on which the boundary between Costa Rica and Nicaragua has been fixed and discussed. The key features of the treaty include Nicaragua's renouncement of all territorial claims over Nicoya in exchange for total Nicaraguan sovereignty over the southern shores of Lake Nicaragua and the entire course of the San Juan River, providing for its free use for trade purposes by Costa Rica (Sibaja 1974; Peralta 1882; Murillo 1986). The boundary thus fixed runs from Salinas Bay on the Pacific Coast, in straight lines, three miles south of the shores of Lake Nicaragua and three miles to the south of the San Juan River up to a point three miles downriver from El Castillo. From here it follows the southern bank of the San Juan River down to the delta. At the San Juan River Delta, the boundary is formed by the southern bank of its northernmost branch, the San Juanillo River, up to the Bay of San Juan del Norte, to a coastal sand bar called Punta Castilla, which constitutes the easternmost marker of the entire boundary (see Figure 7.4).

The Cañas–Jeréz Treaty produced what to this day constitutes an exception in boundary jurisprudence. The Nicaraguan control over the entire course of the San Juan River means that the boundary line runs along its southern bank and not along the thalweg, thus depriving Costa Rica of any claim over canal rights, even though navigation rights for Costa Rican ships were maintained. The dialectics between the boundary disputes and canal surveys are a key indicator of the determining influence the canal issue had on defining and conditioning the boundary between Costa Rica and Nicaragua.

Exclusive Nicaraguan sovereignty over the waters of the San Juan also meant exclusive Nicaraguan participation in any future canal negotiations. The intimate relationship between the fixing of the

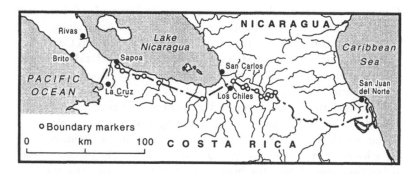

Figure 7.4 Costa Rica–Nicaragua boundary transect

boundary and the potential use of the San Juan River as an interoceanic canal makes this case even more exceptional. The need for maintaining any future canal under the jurisdiction of a single nation dictated in great part the outcome of the boundary delineation. The trade-off obtained by Costa Rica in renouncing any claims to the potential canal in exchange for total sovereignty over Nicoya, has to this day proved a wise choice. The canal has never been built, while Nicoya constitutes a prosperous agricultural region essential to Costa Rica's economy.

The Cañas–Jeréz Treaty brought about a twelve-year lull in boundary disputes between Nicaragua and Costa Rica. The multiplication of canal surveys and feasibility studies conducted along the San Juan River during the last quarter of the nineteenth century served only to renew tensions between the neighbouring countries. A clear manifestation of the dialectics between the boundary and the canal issue is the close coincidence that exists between heightened interest in the canal and major boundary disputes. Every major canal survey along the San Juan has produced a major diplomatic crisis between Nicaragua and Costa Rica. As a corollary of the rivalry existing between major world powers at the time, Nicaragua and Costa Rica competed for the control and potential use of the transboundary watershed.

The decade of the 1870s brought about renewed tensions between Costa Rica and Nicaragua. In 1868, Nicaraguan President Tomás Ayón signed a canal concession with a Frenchman, Michel Chevalier. According to the terms of the Cañas–Jeréz Treaty, Costa Rica was to be consulted prior to any canal concession. This was done and the contract was ratified by the Costa Rican legislature in 1870, when Tomás Guardia seized power in Costa Rica and invalidated the Ayón–Chevalier Concession.[14] In retaliation, Nicaragua repudiated the

99

Cañas–Jeréz Treaty, thus initiating a second round of boundary disputes which were to last until the 1888 Cleveland Arbitration. A special conference on the boundary litigation between Costa Rica and Nicaragua was held in Managua in April 1872. Each side presented divergent interpretations of boundary jurisprudence. Nicaragua claimed that the boundary should run three miles south of the San Juan River, and defined the Sapoá River as the western reference point for the boundary; Nicaragua also insisted on the right of resource exploitation as far south as fifty miles from the San Juan River (Ministerio de Relaciones Exteriores, Nicaragua 1872: 79). The Costa Rican delegation argued that the boundary should run along the southern bank of the lower portion of the San Juan, and identified the Colorado River, which forms the southern branch of the San Juan Delta, as the eastern point of reference for the boundary. Though the 1872 conference failed, it did open the way for arbitration by a third party to resolve the differences. It is interesting to note that the 1872 boundary dispute between Costa Rica and Nicaragua coincided with the canal survey commissioned by the US government led by E.P. Lull. It became clear, following the 1872 conference, that the canal issue was at the crux of the boundary dispute. Nicaragua insisted on total sovereignty over a future canal, while Costa Rica claimed the right to be consulted. Relations between the two countries deteriorated further and diplomatic ties were severed in 1876 (Sibaja 1974).

During the 1880s, canal surveys continued along the San Juan River, particularly with the parties led by Menocal between 1885 and 1886. This fact confirmed the need for an arbitrated settlement, obtained from US President Grover Cleveland in 1888. The Cleveland arbitration reconfirmed the validity of the Cañas–Jeréz Treaty, upholding Costa Rica's rights to navigation on the San Juan River. While Nicaragua was guaranteed exclusive rights over any canal concession, the arbitration also provided for the right of Costa Rica to be consulted prior to any canal concession, and to oppose the deviation of the San Juan River for canalization purposes. The arbitration also identified the need for a precise demarcation of the boundary so as to avoid future discrepancies (Murillo 1986: 52–3).

The boundary itself was demarcated at the turn of the century, following the arbitration by US engineer E.P. Alexander, appointed by President Roosevelt at the request of both countries, between 1897 and 1900. Alexander submitted to Roosevelt, via the official boundary delegations of the two countries, five points of litigation concerning the precise location of the boundary. These were (a) the easternmost point, at Punta Castilla, near San Juan del Norte, (b) along the lower course of the San Juan between the Delta and El Castillo, (c) the southern bank of the San Juan proper, (d) in the area south of Lake Nicaragua, and (e) the westernmost section of the boundary from the Sapoá River to Salinas Bay (Ministerio de Relaciones Exteriores, Nicaragua 1954) (see Figure 7.4).

The arbitration combined a legalistic interpretation of the arguments presented by both parties and an extremely detailed survey of the boundary transect itself. This in turn allowed for a careful location of boundary markers, particularly at points considered potentially sensitive. The demarcation of the boundary proper was completed in 1900, and the results of the Alexander arbitration were ratified by both congresses soon after. The arbitration decisions proposed by E.P. Alexander coincided exactly with the Isthmian Canal Commission's work on the technical feasibility of a canal across the San Juan. Once again, the dialectics between canal and boundary are clearly illustrated. Aside from gathering technical data for engineering decisions, the settling of the boundary dispute over the future canal route was paramount to the US Government's final choice of route. The verdict of the ICC was submitted to the US Congress in 1901, when the Nicaragua route was still seen as the most practicable alternative for a canal.

The events of 1903 were, curiously, destined to tip the scale in favour of Panama. Nonetheless, it is interesting to note the recurrent coincidence between the international interest in the canal and heightened boundary litigations between Costa Rica and Nicaragua. The fact that both countries resorted to arbitration by US Government officials to settle conflicting claims is in a sense ironic, since this was perhaps the country most interested in maintaining the canal route under a single jurisdiction. This fact is clearly reflected in the outcome of both the Cleveland Arbitration of 1888 and the Alexander Arbitration of 1899, both of which reconfirmed Nicaraguan control over the route.

HYDROPOLITICS AND THE GEOPOLITICS OF THE CENTRAL AMERICAN CANAL

The case study of the boundary between Costa Rica and Nicaragua offers an opportunity to reflect on broader issues related to the geopolitics of the Central American isthmus. The control over a potential interoceanic canal route across Central America constituted perhaps the most enduring item of US foreign policy during the second half of the nineteenth century. With the secession of Panama and the completion of the canal in 1914, the Canal Zone became an essential component of the US strategic deployment in the hemisphere during the twentieth century (Lafaber 1978; Sandner 1985; Melville 1898). As Commander Miles Du Val succinctly puts it: 'The Isthmus is the strategic centre of the Americas. The power in possession of it can operate its fleet so as to control both coasts as long as there is uninterrupted transit' (Du Val 1940: 454). In this sense, the Central American isthmus is depicted as the geographical 'pivot' of the American continent, a clear reference to the Anglo-Saxon school of geopolitical thought. One could therefore reshape Mackinder's famous dictum:

Who rules eastern Europe, commands the Rimland
Who rules the Rimland, commands the World Island
Who rules the World Island, commands the world.

as follows:

Who rules the canal route, commands the isthmus
Who rules the isthmus, commands the hemisphere
Who rules the hemisphere, commands the world.

Beyond the whimsical nature of such an exercise in rhetoric, the idea that lies behind this American version of Mackinder's dictum is that the Canal imperative has indeed constituted one of the main bulwarks of US strategy towards Central America. Far more important than its mineral, demographic or economic resources, Central America's strategic location has determined most of the hegemonic rivalries that have marked its history.

In this particular context, how have these geopolitical imperatives conditioned the political destiny of the region and its territorial make-up? The comparative advantage of their isthmian location has spurred Central American states to offer their territories and resources for a potential canal. This fact has fostered endless diplomatic rivalries and intrigues between neighbouring states with comparable isthmian advantages. These rivalries were in turn often manipulated by world powers to serve their geopolitical interests. Whereas the Panama and Nicaragua canal routes constitute in many ways exceptions in boundary jurisprudence, because of their unique geographic attributes, competition for the control of offshore islands and strategic sea lanes involved many Central American states in conflicts with outside powers. Such is the case of the enduring conflict between Colombia and Nicaragua over the Caribbean islands of Providence and San Andrés. In addition to the Panama Canal Zone, the USA controlled several banks and cays (shallow islands) appropriated under the Guano Islands Law of 1856 (Sandner 1985). These Caribbean insular possessions acquired particular strategic significance for their location along key shipping lanes between Panama and US ports. In this sense, the importance of the maritime dimension of Central American boundary conflicts cannot be overlooked, as Sandner (1987) has clearly demonstrated.

The best examples of the correspondence between local and global politics surrounding the canal issue can be found in Panama and Nicaragua during the twentieth century. While the Panama Canal was completed in 1914, the issue of potential interoceanic routes did not disappear altogether. The cyclical resurgence of interest in the Nicaraguan option often responded to particular geopolitical contexts. For instance, the 1931 survey of the Nicaraguan canal route, led by Dan I. Sultan of the US Army Corps of Engineers, coincided with the twelve-year-long military occupation of Nicaragua by US troops (Sultan 1932). The Nicaraguan option has consistently played a role as a potential alternative and, in effect, as a potential bargaining chip in US–Panama relations.

The post-World War II era offers numerous examples of this tendency. A major diplomatic conflict emerged in 1947, when the Panamanian National Assembly rejected the US Postwar Defense Bases Pact which would have extended US leases on military installations outside the Canal Zone. This policy of territorial expansion was greeted by massive anti-US demonstrations (Liss 1978; McKenney 1983). In Nicaragua, the recently installed military regime led by Anastacio Somoza García had not been officially recognized by the US State Department, which even withheld economic and military aid. In response, Somoza deployed a media campaign offering the USA all the necessary facilities for the construction and defense of a canal in Nicaragua (Leonard 1991). When the diplomatic crisis between Panama and the USA subsided with the withdrawal of US troops from outside the Canal Zone, the Somoza regime gained official US recognition, giving Somoza the legitimacy he sought.

A few years later, in a remarkable illustration of linkage politics, the Suez crisis shook the world when Egypt's President Nasser nationalized the canal in 1956. The ramifications were clearly felt in Central America. The Eisenhower administration spared no effort to dissociate Panama from the ongoing crisis. One of these efforts consisted in feigning interest in a Nicaraguan canal so as to deter Panamanian nationalists from following Nasser's example. The tactic was to send a surveying party to Nicaragua

> to impress Panama with the possibility of exercising our option to construct a canal in Nicaragua in order to discourage possible moves by Panama, inspired by developments at Suez, to challenge our Treaty rights in the Canal Zone, also to bring about a more reasonable attitude on the part of Panama on Canal Zone problems.
>
> (Rubottom 1956)

While the planned survey was aborted owing to Somoza García's assassination, it sheds considerable light on the specific weight given by the US Government to the Nicaraguan option as a bargaining chip against Panama. The issue has resurfaced regularly, as the 1964, 1968 and 1989 events in Panama have shown, fuelling once again speculation over alternatives canal routes. The US control over the Panama Canal Zone has been the object of constant resentment by Panamanian nationalists. The Carter–Torrijos Treaty of 1977 was aimed at assuaging these tensions, and the key feature of this treaty was precisely the abolition of the Canal Zone as an extra-territorial enclave and its return to Panamanian sovereignty by 1999.

The Nicaragua option has not been under strict US monopoly. Under the Sandinista Government, in a clear defiance of US hegemony in the region, Japanese industrialists were solicited in 1989 to design and fund a sea-level canal across Nicaragua, a situation similar to that of eighty years earlier when Nicaragua's liberal President Zelaya flirted with Prussian interests offering conditions for a canal. In the latter instance, the result was the 1909 invasion of Nicaragua by US troops to secure exclusive control over the Panama canal, then under construction (Bermann 1986).

These examples shed light on the fundamental role played by the canal issue in Central American politics. Because of its explicit territorial content, the issue of an interoceanic waterway under hegemonic control has consistently led to local and global rivalries. Perhaps the most salient feature of this analysis is the recurrent nature of the canal imperative. Despite the existence of the Panama Canal, other geographical and technical options take on renewed importance at critical periods in history. Generally, these are associated with periods of hegemonic transition (Britain–US, US–Japan). It is no coincidence that Japan and the USA constitute, with the Panamanian Government, the Tripartite Commission for the interoceanic canal. The history of the canal issue has reflected the main episodes of the world economy and the changing relationships of power at a global scale.

CONCLUSION

There have been complex interactions between territory, boundaries and water resources along potential canal routes in the Central American isthmus. The case studies of Nicaragua and Panama clearly exemplify the intimate linkages between global geopolitics and local politics. The political destiny of these two countries has perhaps been most dramatically influenced by their geographic location. In this sense, the transisthmian imperative seems to have been more influential in dividing the isthmus than in unifying it. One could even assert that it has constituted one of the principal opposing forces to Central American political integration.

While the concept of hydropolitics has been coined principally for describing conflict between states over shared water resources, especially in the Middle East, its application to the canal issue in Central America provides a convenient conceptual framework within which to compare boundary situations in other regions of the world. While most conflicts over water resources imply their use for agricultural or energy–related projects, in this case control over the canal route centred on the waterway itself. The case studies of Nicaragua and Panama show how closely associated the issues of water resources and rights to the canal have been historically in the boundary disputes.

However, technological changes may be spelling an end to the traditional vision of a ship canal across Central America. The rapid evolution of transport technology, especially in terms of high-speed land bridges and non-vehicular transport systems (liquid and slurry pipelines), is transforming the absolute advantage of Central America's isthmian geography. Recent initiatives in Costa Rica and Panama clearly confirm the growing trend towards the promotion of 'dry canals', involving pipelines or high-speed trains, across the isthmus, thus relegating the classic engineering puzzles associated with water control in traditional ship canals to a thing of the past. While these dry canals are usually capital-intensive, they have the advantage of being more easily administered, reducing to a minimum the liabilities associated with the management of large watersheds to feed a ship-canal.

Future boundary conflicts in Central America will therefore not necessarily involve water resources or conflicts over canal routes as they have in the past. Maritime boundaries and conflicts over transboundary exploitation of natural resources are more likely to be the rule. The emergence of several transboundary environmental initiatives such as the La Amistad Biosphere Reserve between Panama and Costa Rica and the SIAPAZ project between Costa Rica and Nicaragua appear as a new trend in regional environmental integration. These initiatives are bound to create new perspectives and problems for the region. As a potential opposing force to the transisthmian prerogative, this recent trend in transboundary parks may signal a new and fascinating chapter in territorial relations in Central America.

NOTES

1 This research was made possible by a collaborative grant from the MacArthur Foundation's Program for Peace and International Cooperation for 1990–1991. The co-grantee was Professor Bernard Nietschmann from the Department of Geography, University of California at Berkeley. The author wishes to thank several persons who made this contribution possible. In the first place, Christian Brannstromm, Graduate Student at the University of Wisconsin, provided me with keen insight and outstanding material from his research in the National Archives in Washington, the US Library of Congress, and the National Archives of Costa Rica and Nicaragua. Carolyn Hall, Professor at the University of Costa Rica's Geography Department, gave careful reading to earlier drafts of this chapter for which I am extremely grateful. Finally, I wish to thank Eduardo Rodríguez for the draughting of the maps illustrating this piece.

2 See Castillero (1954) on the history of interoceanic communications, also Mack (1944) provides a good synopsis of the canal imperative in Central America.

3 The Caribbean at the onset of the Spanish conquest, centred around what is today the Dominican Republic, Haiti and Cuba, is referred to by scholars as the Spanish Main.

4 See A. Humbolt (1811), also a good appraisal of Humboldt's contribution to canal studies can be found in McCullough (1977), Castillero (1954) and Mack (1944).

5 Selser (1982: 53) and Castillero (1954) for reference on Bolivar and the Central American canal.

6 McCullough (1977) deals extensively with the French venture in Panama.

7 Canal concessions in Panama are dealt with in detail by Castillero (1954), Selser (1982), Lafaber (1978) and Mack (1944).

8 See Selser (1982: 211–50) for a detailed account of the particular political concessions explicit in the Hay–Bunau Varilla Treaty.

9 For more details on US–Panamanian relations and the canal issue see Liss (1978), McKenney (1983), Lafaber (1978) and Ealy (1971).

10 On Anglo–Spanish rivalry in Central America see Floyd (1967), and concerning the Mosquito Coast see Dozier (1985).

11 For the report of the ICC, see U.S. Congress, Senate (1904); McCullough

(1977) and G. Selser (1982) also deal extensively with the comparative advantages of the Nicaraguan and Panamanian option at the turn of the century.

12 Dozier (1985) provides a detailed account of the rise and fall of interest in the canal in the San Juan River's history.

13 Sibaja (1974: 148); also see Peralta (1882) and Peralta (1898).

14 For more information concerning the boundary litigations between Costa Rica and Nicaragua during the 1870s, see Ayón (1872), Sibaja (1974) and Murillo (1986).

BIBLIOGRAPHY

Atkins, T.B. (1890) *Report on the Tonnage of Traffic within the Zone of Attraction for the Maritime Canal of Nicaragua in 1890 and Estimated for 1897*, New York: New York Print Co.

—— (1891) *The Inter-Oceanic Canal Across Nicaragua and the Attitude of the Government of the U.S.*, New York: New York Print Co.

Ayón, T. (1872) *La cuestion de límites territoriales entre las Repúblicas de Nicaragua y Costa Rica*, Managua: Imprenta del 'Centroamericano'.

Belly, F. (1867) *A travers l'Amérique Centrale: Le Nicaragua et le Canal Interocéanique*. vol.1, Paris: Imprimerie Victor Goupy.

Bermann, K. (1986) *Under the Big Stick: Nicaragua and the United States Since 1848*, Boston: South End Press.

Castillero, R.E. (1954) *Historia de la Comunicación Interoceánica y de su influencia en la formación y en el desarrollo de la Entidad Nacional Panameña*, Panama: Imprenta Nacional.

Childs, O.W. (1851) *Nicaragua Canal: Map and Profile of the Route for Construction of the Shipcanal*, New York: W.C. Bryant and Co. Printers.

Costa Rica Comisión De Límites (1987) *Exposición presentada por la comisión de límites de Costa Rica al ingeniero arbitro honorable E.P. Alexander, el día 30 de junio de 1897*, San José: Tipografía Nacional.

Dozier, C.L. (1985) *Nicaragua's Mosquito Shore: The Years of British and American Presence*, Birmingham: University of Alabama Press.

Dutton, C.E. (1892) *The Nicaragua Canal*, Washington, DC: U.S. Congress, Senate, 52nd Congress, 1st session, Miscellaneous Document no. 97.

Du Val, M.P. (1940) *Cadiz to Cathay: The Story of the Long Struggle for a Waterway Across the American Isthmus*, Stanford: Stanford University Press.

Ealy, L.O. (1971) *Yanqui Politics and the Isthmian Canal*, University Park: Pennsylvannia State University.

Floyd, T.S. (1967) *The Anglo–Spanish Struggle for Mosquitia*, Albuquerque: University of New Mexico Press.

Folkman, D.I. (1972) *The Nicaragua Route*, Salt Lake City: University of Utah Press.

Girot, P.O. (1989) 'Formación y estructuración de una Frontera Viva: El caso de la región Norte de Costa Rica', *Geoistmo*, vol. III, no. 2: 17–42.

Gonzalez, P. (1976) 'La ruta de Sarapiquí: Historia socio-política de un camino', *Avances de Investigación*, research paper no. 15, San Jose: Universidad de Costa Rica.

Holland to Harrington (1956) September 6, Washington, D.C., *FRUS 1955–57*, Washington, DC: National Archives, 304–5.

von Humboldt, A. (1811) *Political Essay on the Kingdom of New Spain*, London: Longman.

Ireland, G. (1941) *Boundaries, Possessions and Conflicts in Central and North America and the Caribbean*, Cambridge, MA: Harvard University Press.

Keasbey, L.M. (1896) *The Nicaragua Canal and the Monroe Doctrine*, New York: G.P. Putnam's Sons.

Lafaber, W. (1978) *The Panama Canal: The Crisis in Historical Perspective*, New York: Oxford University Press.

Leonard, T.M. (1991) *Central America and the United States: The Search for Stability*, Athens, GA: University of Georgia Press.

Liss, S.B. (1978) *The Canal Aspects of United States–Panama Relations*, Notre Dame: Notre Dame University Press.

Loos, E. (1863) *Empresa Centro-Americana y Universal del Canal de Nicaragua*, Managua: Imprenta del Gobierno.

Mack, G. (1944) *The Land Divided: A History of the Panama Canal and Other Isthmian Canal Projects*, New York: Alfred A. Knopf.

Maritime Canal Company of Nicaragua (1889) *Nicaragua, the Gateway to the Pacific*, New York: J. Bien and Co.

—— (1890) *The Maritime Ship Canal of Nicaragua*, New York: The Maritime Canal Company of Nicaragua.

McCullough, D. (1977) *The Path Between the Seas: The Creation of the Panama Canal, 1870–1914*, New York: Simon & Schuster.

McKenney, J.W. (1983) *US–Panama Relations 1903–1978: A Study in Linkage Politics*, Boulder, CO: Westview Press.

Melville, G.W. (1898) *The Strategic and Commercial Value of the Nicaragua Canal, the Future Control of the Pacific*, Washington DC: Government Printing Office.

Menocal, A.G. (1890) *The Nicaragua Canal: Its Design, Final Location and Work Accomplished*, New York: Press of the New York Printing Co.

Ministerio de Relaciones Exteriores, Nicaragua (1872) *Documentos relativos a las últimas negociaciones entre Nicaragua y Costa Rica sobre límites territoriales y canal interoceánico*, Managua: Imprenta Nacional.

—— (1954) *Situación jurídica del Río San Juan*, Managua: Imprenta Nacional.

Molina, F. (1851) *Memoir on the Boundary Question Pending Between the Republic of Costa Rica and the State of Nicaragua*, Washington, DC: Gideon and Co.

Murillo J.H. (1986) 'La controversia de límites entre Costa Rica y Nicaragua: el laudo Cleveland y los derechos canaleros 1821–1903', *Anuario de Estudios Centroamericanos*, 12, 2: 45–58.

Peralta, M.M. (1882) *El Río San Juan de Nicaragua, derechos de sus rebereños, las Repúblicas de Costa Rica y Nicaragua ... según documentos históricos*, Madrid: M. Murillo.

—— (1898) *Costa Rica y costa Mosquitos: Documentos Para la Historia de la Jurisdicción Territorial de Costa Rica y Colombia*, Madrid: M. Murillo.

Rubottom to Richards, September 19, 1956, Washington, DC, *FRUS 1955–57*, 305–6.

Salisbury, R.V. (1975) *Costa Rican Relations with Central America, 1900–1934*, Buffalo: Council on International Studies.
Sandner, G. (1985) *Zentralamerika un der ferne karibische Westen: Konjunkturen, Krisen un Konflict 1503–1984*, Stuttgart: F. Steiner Wiesbaden Gesellshaft.
—— (1987) 'Aspectos de problemas de geografía 'Territorial' del Mar Caribe en le contexto de las nuevas delimitaciones', *GEOISTMO*, 1, 1: 9–32.
Selser, G. (1982) *El Rapto de Panamá*, San Jose: Editorial Universitario Centroamericano.
Sibaja, L.F. (1974) *Nuestro límite con Nicaragua: Estudio histórico*, San Jose: Don Bosco.
Sultan, D.I. (1932) 'An army engineer explores Nicaragua', *National Geographic Magazine* 61: 593–627.
U.S. Congress, Senate, House Committee on Foreign Affairs (1882) *Nicaragua Canal Report, to accompany H.R. 6799*, 47th Congress, House Report no. 1698, Washington, DC: Government Printing Office.
U.S. Congress, Senate, (1904) *Report of the Isthmian Canal Commission, 1899–1901*, 58th Congress, 2nd Session, Doc. no. 222, Washington, DC: Government Printing Office.
U.S. Congress, Senate, Navy Department (1874) *Reports of Explorations and Surveys for the Location of a Ship Canal between the Atlantic and Pacific Oceans*, 43rd Congress, 1st Session, Senate Ex. Doc. no. 57, Washington, DC: Government Printing Office.
—— (1897) *Reports of Explorations and Surveys*, 45th Congress, 3rd session, Senate Ex. Doc. no. 75, Washington, DC: Government Printing Office.
—— (1886) *Report of the U.S. Nicaragua Surveying Party 1885 by A.G. Menocal*, Washington, DC: Government Printing Office.
U.S. Nicaragua Canal Board (1896) *Report of the Nicaragua Canal Board*, Washington DC: Government Printing Office.
U.S. Nicaragua Canal Commission (1899) *Report of the Nicaragua Canal Commission, 1897–1899*, Baltimore: The Frienwald Co.
Whelan to Rubottom (1958) June 5, Managua, Central Decimal Files 1955–59, RG 59, NARA.
Williams, M.H. (1971) 'The San Juan River–Lake Nicaragua Waterway, 1502–1921.' Ph.D. dissertation, Louisiana State University.

Part III

BOUNDARIES IN SOUTH AMERICA

8

THE ECUADOR–PERU DISPUTE

A reconsideration

Ronald Bruce St John

INTRODUCTION

At the outset of the independence era, the exact borders of the nascent republics of Latin America were a highly controversial subject. As a result, bitter territorial disputes, often involving vast tracts of land and considerable wealth, quickly developed. Many of these territorial questions were in fact boundary disputes which resulted from Spain's failure to delineate carefully its administrative units during the colonial period. The boundary dispute between Ecuador and Peru, sometimes referred to as the Zarumilla–Marañón dispute, falls into this category. Other irredentist issues between South American neighbouring states arose from challenging the validity of treaty settlements previously ratified by the parties to a dispute. An example of the latter was the Tacna–Arica question which plagued relations between Chile and Peru for over four decades after the end of the War of the Pacific. Emotionally charged and highly involved, territorial issues have complicated and disrupted inter-American relations throughout the nineteenth and twentieth centuries.

THE GEOGRAPHICAL CONTEXT

The boundary dispute between Ecuador and Peru actually involved the three distinct territories of Tumbes, Jaén, and Maynas (Ministerio de Relaciones Exteriores, Peru 1961). Tumbes is an extremely arid region of some 500 square miles located on the Pacific seaboard between the Tumbes and Zarumilla Rivers. Jaén is an area of close to 4,000 square miles which lies on the eastern side of the Andean Range between the Chinchipe and Huancabamba Rivers which drain into the vast Amazon

113

Basin. Both Tumbes and Jaén were subject to Peruvian sovereignty after 1821, the year Peru declared independence from Spain, and delegates from both areas attended the Peruvian congresses held in 1822, 1826, and 1827 (Wagner de Reyna 1962: 41; Maier 1969: 28–9). Maynas, often referred to in Ecuador as the *Oriente*, is the third and largest of the disputed territories, covering well over 100,000 square miles of land. Triangularly shaped, the limits of the region are bounded by the headwaters of the Amazon tributaries to the west, the Yapurá and

Figure 8.1 Ecuador's Amazonian boundaries
Source: Deler, J.-P. 1981

114

Caquetá Rivers to the north, and the Chinchipe, Marañón and Amazon Rivers to the south (see Figure 8.1). Maynas was liberated from Spanish rule in 1821, and true independence was achieved by 1822. Representatives from Maynas attended the 1826 and 1827 Peruvian congresses. After independence, Peru occupied much more of the vast area of Maynas than Ecuador; but the inhospitable character of the terrain hampered either party's ability to exert effective jurisdiction (Wright 1941: 253–4).

THE HISTORICAL BACKGROUND

The conflicting claims of Ecuador and Peru arose from the uncertainty of Spanish colonial administrative and territorial divisions. The Crown made little effort to delimit carefully the boundaries of its possessions because most of those boundaries lay in remote and sparsely inhabited areas of marginal importance to the colonial economy. As a result, colonial jurisdictions were often vague and overlapping while boundary surveys were either inadequate or nonexistent. With the advent of independence, boundary issues assumed a new importance because they now became questions of territorial possession previously under a single realm. Consequently, even when neighbouring republics agreed that their new national borders should reflect those of the former colonial administrative units, they still confronted great difficulties in delineating their boundaries. To complicate matters further, the wars of independence generated or accentuated personal and regional jealousies; and these rivalries hardened as states vied for political and economic advantages. Separatist sentiments in many states, particularly in the south of Peru, added yet another element of discord. In this sense, the Ecuador–Peru dispute was typical of the many boundary disputes which complicated diplomatic relations in post-independence Latin America.

The overriding importance given to territorial questions led many Latin American governments swiftly to assert their rights to disputed regions. On July 6, 1822, Bernardo Monteagudo, Peruvian Minister of War and Marine, and Joaquín Mosquera, the Colombian Ambassador to Peru, called for a precise demarcation of limits at an unspecified later date. An article in the 1823 Peruvian constitution called for the Peruvian Congress to fix the boundaries of the republic; and on February 17, 1825, Foreign Minister José Faustino Sánchez Carrión again asked congress to resolve the nation's borders. In the face of such appeals, the Peruvian Congress appointed a boundary commission, but the political and economic uncertainty of the times made any real progress impossible (Pérez Concha 1961: vol. I, 53–7; Basadre 1968: vol. I, 67–9, 203–6).

During the struggle for independence, Ecuador and Peru had joined other Latin American states in accepting the doctrine of *uti possidetis de jure* as the principal method of establishing the boundaries of newly independent states. Under this principle of international law, Latin American states as former Spanish colonies agreed that each new state was established from the territory of the colonial administrative areas. In the case of Peru, for example, this meant that the limits of the new republic would be defined by the previous jurisdiction of the Viceroyalty of Peru, the *Audiencia* of Lima, and the *Audiencia* of Cuzco. While the concept of *uti possidetis* was generally accepted throughout Latin America, the doctrine was of questionable validity under international law and proved extremely difficult to apply. Colonial documents were complex, and the language which the Spanish Crown employed to make territorial changes often lacked clarity. As a result, confusing and sometimes contradictory legal bases were often the only foundation for significant reforms to the colonial administrative system (Checa Drouet 1936: 137–8).

THE CRUX OF THE BOUNDARY LITIGATION

The Ecuadorian Government based its legal case for the application of *uti possidetis* on a series of Spanish decrees issued after 1563 when a *cédula* awarded Maynas, Quijos, Jaén, and adjoining land, i.e., the whole of the disputed territory, to the *Audiencia* of Quito. Based on the doctrine of *uti possidetis* and the *cédulas* of 1563, 1717, 1739 and 1740, Ecuador argued that the disputed territories were first part of the *Audiencia* of Quito, later part of Greater Colombia, and finally part of Ecuador when the latter emerged in 1830 following the breakup of Greater Colombia (Flores 1921: 67–70; Delgado 1939: 44–53). In turn, the Peruvian Government argued that the essence of independence in the Americas was the sacred and unalterable character of self-determination movements. Within this greater principle, Peru contended that *uti possidetis* served only as a guide for the demarcation of actual boundaries and not as a basic principle for the assignment of provinces or the organization of states (Ministerio de Relaciones Exteriores, Peru 1937: 3, 14; Tudela 1941: 12–38). This aspect of the Peruvian legal case was based on a recognized corollary to the rule of *uti possidetis* which gave individual provinces the right to attach themselves to the state of their choosing. Following this line of argument, the

Peruvian Government concluded that all of the territories in question were Peruvian because the populations of Jaén, Tumbes and Maynas had all voluntarily adhered to Peru at the time of Peruvian independence, well before the independence of Ecuador (Cornejo and de Osma 1909: 16–17; Porras Barrenechea 1942: 7).

In support of this chief argument that the principle of self-determination was the most relevant to the ownership question, the Peruvian Government developed two main arguments. Through a *cédula* dated July 15, 1802, the King of Spain had separated the provinces of Maynas and Quijos, excluding Papallacta, from the Viceroyalty of New Granada, for ecclesiastical and military purposes, and transferred them to the Viceroyalty of Peru. The Peruvian Government claimed that the 1802 *cédula* was also a valid guide for determining the jurisdiction of Maynas; however, it was always careful to put forward this claim as secondary to its title based on the principle of self-determination. Pressing for the applicability of the older colonial decrees, the Ecuadorian Government sought to counter this Peruvian argument by contending that the 1802 *cédula* separated Maynas and Quijos for ecclesiastical and administrative ends but not in any political sense (Wagner de Reyna 1964: vol. I, 8–9; Zook 1964: 28–30).[1] In addition, the Peruvian Government argued that the principle of *uti possidetis* was not applicable until the end of colonial dependence which it interpreted to be the 1824 Battle of Ayacucho. Since 1810 was widely accepted throughout Latin America as the year in which *uti possidetis* was applicable, the Ecuadorian Government naturally refused to accept the latter date, especially since by that time the populations of Jaén, Tumbes and Maynas had all expressed their determination to become part of Peru (Santamaría de Paredes 1910: 277–80; Ulloa Sotomayor 1941: 19–20).[2]

Other documents of legal importance to the dispute included the treaties of 1829 and 1832 and the highly controversial Pedemonte–Mosquera Protocol of 1830. In the wake of an abortive Peruvian invasion of Ecuador, the two governments concluded on September 22, 1829, a peace treaty known as the Larrea–Gual Treaty (Zook 1964: 271–9). The 1829 agreement was a general instrument of peace and not exclusively one of frontiers. While it recognized as the boundary between the signatories the limits of the ancient Viceroyalties of New Granada and Peru, it neither settled the boundary question nor fixed a boundary line. The pact did not even mention Jaén, Tumbes or Maynas, much less impose on Peru a specific obligation to surrender those territories (see Figure 8.1). It merely established a settlement procedure to be followed. Article VI of the treaty left the final solution to a commission of limits which was to meet within forty days of treaty ratification and complete its work within six months. Treaty ratifications were exchanged on October 27, 1829, but the assent of Greater Colombia was of doubtful validity as it was ratified without congressional approval. Boundary negotiations between Greater Colombia and Peru were subsequently halted in May 1830, when the former split into three secessionist states. Thereafter, the Peruvian government refused to be bound by the terms of the Larrea–Gual Treaty (Pérez Concha 1961: vol. I, 78–86; Wagner de Reyna 1964: vol. I, 25).[3]

Some two years later, on July 12, 1832, the governments of Peru and Ecuador concluded a treaty of friendship, alliance and commerce in which they agreed to recognize and respect their present limits until a boundary convention could be negotiated. Unfortunately, the terms of the treaty did not specify whether the phrase 'present limits' referred to the territories then in the physical possession of the signatories or to the territories of the former viceroyalties mentioned in 1829. The Peruvian Government, arguing that the 1832 treaty nullified the 1829 pact, gravitated towards the first interpretation while the Ecuadorian Government, arguing that the 1832 treaty confirmed the 1829 treaty, advocated the second. Valid ratifications of the 1832 treaty were exchanged on December 27, 1832 (Cano 1925: 48; Eguiguren 1941: 149; Zook 1964: 282–5).

THE PEDEMONTE–MOSQUERA PROTOCOL

The ensuing debate between Ecuador and Peru over the relevance of the 1829 and 1832 treaties involved several complicated issues. On the one hand, there was the question of the extent to which the 1829 treaty actually established a boundary. While Peru argued that the pact established only a principle of delimitation and a procedure to be followed, Ecuador maintained that the treaty actually fixed a boundary and thus resolved the controversy. In support of its position, the Ecuadorian Government later introduced the Pedemonte–Mosquera Protocol into its legal brief. According to Ecuador, the Peruvian Foreign Minister, Carlos Pedemonte, and the Greater Colombian Minister to Peru, General Tomás C. Mosquera, agreed to a protocol on August 11, 1830, which determined the bases of departure for the border commissioners established in the 1829 treaty. In this protocol, Foreign Minister Pedemonte supposedly accepted the Marañón River as the frontier between Peru and Ecuador, leaving in doubt only the question

118

of whether the border would be completed with the Chinchipe or Huancabamba Rivers. The Colombian Government was long in possession of a copy of the Pedemonte–Mosquera Protocol but did not mention it until 1904, and the Ecuadorian government first introduced the document in an *exposición* filed on October 20, 1906. The Peruvian Government rejected both the validity and applicability of the protocol, and in support of its position it demonstrated that General Mosquera had sailed from the port of Callao on the day before the protocol was allegedly concluded. Even if General Mosquera had reached an agreement on August 11, 1830, Peruvians pointed out that he could not at that time be considered an official representative of Greater Colombia because Venezuela had seceded at an earlier date which meant Greater Colombia had ceased to exist as a legal entity. Finally, the Peruvian Government emphasized that any document of the importance of the Pedemonte–Mosquera Protocol would have required some form of congressional approval and none was given (Ulloa Cisneros 1911; San Cristóval 1932: 43–83; Zook 1964: 279–81).

The second major area of disagreement centred on whether or not Ecuador was entitled to assume the legal privileges and duties of Greater Colombia after the latter disintegrated. Although Ecuador enthusiastically advocated this position, its legal case here was at best questionable. According to the doctrine of the secession of states, when a state ceases to exist, its treaty rights and obligations generally cease with it. Therefore, after Greater Colombia split into three secessionist states in 1830, there was a legitimate question in Peru as well as internationally as to why Ecuador should feel it was the legitimate successor to Greater Colombia. Moreover, even if Ecuador had some limited claim to the legal rights and obligations of Greater Colombia, it could hardly be the successor to the latter's southern boundary since that line had never been fixed. As a point of fact, the boundary commission provided for in the 1829 treaty never met (Brierly 1963: 153–4; Maier 1969: 39).

The third and final issue focused on the exact interrelationship of the 1829 and 1832 agreements. The Peruvian Government took the position that the 1832 treaty both nullified the earlier pact and confirmed Peruvian possession of Jaén, Tumbes and Maynas. In turn, the Ecuadorian Government argued that the 1829 treaty fixed a final boundary which was unaffected by the later agreements. As for the Peruvian argument that the 1832 treaty rendered the 1829 pact null and void, there was certainly no clear statement to this effect in the 1832 agreement. On the other hand, as we have seen, it was far from clear that Ecuador inherited the rights and obligations of the 1839 treaty. Finally, since the 1829 agreement did not establish a boundary, it remained impossible to determine whether the 'present limits' in the 1832 agreement referred to the Viceroyalties of Peru and New Grenada in the 1829 treaty or to those territories in the actual possession of Peru and Ecuador when they concluded the 1832 treaty (Tudela 1941: 12–38; Zook 1964: 23–4; Maier 1969: 40).

For much of the next fifty years, the boundary dispute dominated diplomatic relations between Peru and Ecuador. In late 1839, the Quito Government proposed to Chile an abortive plan which included the cession of northern Peru to Ecuador, and in early 1842, Ecuador threatened to occupy Jaén and Maynas by force if Peru refused to cede them voluntarily. Two decades later, an Ecuadorian attempt to cede to English creditors land claimed by Peru in the Amazon region of Canelos led to a Peruvian invasion of Ecuador. The Treaty of Mapasingue, dated January 25, 1860, ended the Peruvian invasion and re-established diplomatic relations between the two states. In the agreement, the Ecuadorian regime agreed to nullify the cession of Amazonian lands and provisionally to accept Peruvian claims to the disputed territories on the basis of *uti possidetis* and the *cédula* of 1802. At the same time, it reserved the right to present, within two years, new documents in support of its territorial claims. On the other hand, if it failed to present documents annulling Peru's right of ownership within the specified period, it would lose its rights and a mixed commission would fix the border based on Peruvian pretensions. Highly favourable to Peru, the Treaty of Mapasingue proved a pyrrhic victory as a unified Ecuadorian Government later established itself in Quito and declared the 1860 treaty null and void (García Salazar 1928: 112–18 and 134–42; Pérez Concha 1961: vol. I, 109–27 and 152–81).

THE WAR OF THE PACIFIC AND THE ECUADOR–PERU DISPUTE

In the second half of the 1860s, the Ecuador–Peru dispute was temporarily set aside in the face of the Spanish intervention in the Americas; nevertheless, it again surfaced at the end of the following decade. In the build-up to the War of the Pacific, in March 1879 the Chilean Government sent an emissary to Quito with instructions to bring Ecuador into the War of the Pacific on the side of Chile. The Chilean envoy was told to suggest to the Ecuadorian Government that the time was ripe to resolve its dispute with Peru by occupying the contested territory. If Ecuador rejected this proposal, the Chilean diplomat was instructed to negotiate an offensive and defensive alliance. While the Ecuadorian Government eventually chose to remain neutral, regional diplomacy in this period exemplified the extent to which bilateral boundary disputes in Latin America often assumed multilateral dimensions as neighbouring states formed alliances to attain their foreign policy objectives. In the Amazon region, three separate but related disputes over the ownership of the Amazon Basin involved Peru and Ecuador, Peru and Colombia, and Colombia and Ecuador. Towards the end of the nineteenth century, the Chilean Government further complicated matters by encouraging the Amazonian claims of Colombia and Ecuador in a effort to distract Peru from the Tacna–Arica question (Burr 1965: 146–7; Soder 1970: 64–5).

In 1887, the Ecuadorian Government again tried to cancel foreign debts by granting land concessions in a section of the Amazon Basin claimed by Peru. As a result, the two governments opened new negotiations which led on August 1, 1887 to an agreement known as the Espinosa–Bonifaz Convention. Under its terms, the signatories agreed to submit their territorial dispute to an arbitration by the King of Spain. The agreement provided for an arbitration so complete that even the points in contention were left to the arbiter with no principles for their definition specified. Ecuadorian critics of the convention later argued it was null and void because the open-ended procedure offered no securities for the weaker party. The agreement also provided for direct negotiations to continue concurrently with the arbitration process; and if the former were successful, their results would be brought to the knowledge of the arbitrator. Both Ecuador and Peru had more faith in direct negotiations than in the Spanish arbitration, and serious talks aimed at a comprehensive settlement soon produced an agreement (Ministerio de Relaciones Exteriores, Peru 1936a: 271–3 and 1890: 79–80; Tobar Donoso and Tobar 1961: 147–51).

THE SPANISH ARBITRATION

The García–Herrera Treaty, dated May 2, 1890, granted Ecuador extensive concessions in the *Oriente*, including access to the Marañón River from the Santiago River to the Pastaza River. Since the Peruvian Government had long opposed making Ecuador an Amazonian power, the terms of the treaty marked a high watermark of compromise for Peru. Faced with a very favourable settlement, the Ecuadorian Congress quickly approved the pact on July 19, 1890. The Peruvian Congress conditionally approved the treaty on October 25, 1891, but refused to grant final approval until modifications were made to several articles. The changes demanded would have given Peru a much larger share of the disputed territory while restricting Ecuador's Marañón River access to the mouth of the Santiago River. In 1893, the Peruvian Congress reconsidered the terms of the García–Herrera Treaty but continued to insist on either treaty modifications or a full arbitration by the King of Spain. Ecuador refused to accept the Peruvian proposals, and on July 25, 1894, the Ecuadorian Congress revoked its approval of the pact, while directing the government to open new talks (Zook 1964: 295–9; Wood 1978: 3–4).

Several considerations help explain the Peruvian Government's agreement to the García–Herrera Treaty, a pact through which it would have lost some 120,000 square miles of Amazon jungle. First, Peru had not recovered economically or politically from the consequences of the War of the Pacific, and in just four years, it was scheduled to participate in the Tacna–Arica plebiscite. Needing all available resources to protect its southern interests, the García–Herrera Treaty was a means to neutralize Ecuador while Peru focused on its struggle with Chile. Second, the Peruvian Government lacked the necessary legal documents to prove conclusively its ownership of the disputed territories. Peruvian scholars had been searching feverishly for new documents in Seville and other Spanish archives, but they had yet to discover anything which proved decisively the Peruvian case. A third consideration, not always mentioned, related to the relative value of the *Oriente*. When compared to Tacna and Arica, the Amazon territory was geographically larger and of greater potential wealth; but it was also situated in a remote area, less known to Peruvians. Furthermore, it had not been the theatre of long, bloody war. The boom in rubber prices had not yet occurred and at the time little or no thought was given to the possibility of oil deposits in the region. Consequently, there were both economic and political reasons for the Peruvian Government to assign a higher priority to a successful resolution of the Tacna and Arica dispute even if it meant granting concessions in the *Oriente* (Ulloa Sotomayor 1942: 67–71; Wagner de Reyna 1964: vol. I, 34–5).

In 1890 and again in 1891, the Colombian Government protested that the terms of the García–Herrera Treaty violated its territorial rights. Faced with continuing Colombian opposition, the governments of Peru and Ecuador eventually agreed to broaden the 1887 arbitral convention to include Colombia. The Tripartite Additional Arbitration Convention, dated December 15, 1894, provided for Colombian adherence to the arbitration provisions of the 1887 Espinosa–Bonifaz Convention. It also provided for an arbitral decision based on legal title as well as equity and convenience. As it turned out, the tripartite convention never came into effect because the Ecuadorian Congress rejected the pact. Ecuadorian critics rightly feared that a tripartite settlement might lead to Peru and Colombia dividing the *Oriente* between themselves at Ecuador's expense. When it became clear that Ecuador would not ratify the 1894 convention, the Peruvian Congress revoked its approval of the Tripartite Additional Arbitration Convention. This cleared the way for a resumption of the Spanish arbitration and in March 1904 both governments asked the King of Spain to continue this procedure (Pérez Concha 1961: vol. I, 256 and 270–84; Ministerio de Relaciones Exteriores, Peru, 1896: 153–61).

The 1887 Spanish arbitration led to a projected award in 1910 which largely accepted Peru's juridical theses. Rejecting Ecuador's attempt to reconstitute Viceroyalties and *Audiencias* dating back to 1563, the arbitration decision agreed with the central Peruvian argument that the real issue was one of fixing the boundaries between provinces which had chosen at the time of independence to join one state or the other. Accepting the rule of *uti possidetis*, the decision agreed to the validity of the royal *cédula* of 1802 as well as older decreees. As to the documents pivotal to the Ecuadorian case, the arbitration rejected the 1829 treaty on the grounds that Ecuador lost its rights as a successor to Greater Colombia when it concluded the 1832 treaty. It also ruled that the Pedemonte–Mosquera Protocol lacked authenticity as well as the required approval of the Peruvian and Ecuadorian congresses. Finally, the arbitration agreed that the 1832 treaty had been ratified and that the ratifications had been duly exchanged (Flores 1921: 56–62; Ministerio de Relaciones Exteriores, Peru 1936a: 12–18) (see Figure 8.1).

When the provisions of the arbitration became known in Ecuador, they produced violent demonstrations against Peru in Quito and Guayaquil. When news of these riots reached Peru, they led to reprisals in Lima and Callao and both countries assumed a war footing. The mobilization in Peru alone put 23,000 men in arms. The Ecuadorian Government suggested direct negotiations in Washington, but Peru refused to consider any solution other than arbitration. While war appeared imminent, a tripartite mediation by Argentina, Brazil and the United States eventually restored peace. In November 1910, the King of Spain decided not to pronounce his arbitration. With the end of the Spanish arbitration, the mediating powers advised Peru and Ecuador to bring the dispute before the Permanent Court of Arbitration at The

Hague. The Peruvian Government accepted this proposal, but Ecuador continued to insist on direct negotiations (Pérez Concha 1961: vol. I, 341–93; Basadre 1968: 94–102).

In retrospect, the Ecuadorian Government made a serious mistake when it encouraged public demonstrations against the pending Spanish judgement. While the projected award was favourable to Peru, the King of Spain awarded Ecuador much more territory than it was to receive in the final settlement three decades later. Moreover, the abortive arbitration presented Peru with a major diplomatic victory both because of the favourable terms of the projected award and because Ecuador's reaction cast it in a unfavourable light. Finally, in the negotiations with Ecuador, time was on the side of Peru as it remained deadlocked with Chile in the Tacna and Arica dispute. With the breakdown in negotiations, Peru continued in *de facto* control of most of the disputed territory, a control buttressed by a strong deployment of armed force (Ulloa Sotomayor 1942: 68; Zook 1964: 97–108).

THE SALOMON–LOZANO TREATY BETWEEN PERU AND COLOMBIA

On March 24, 1922, the governments of Peru and Colombia concluded a treaty of frontiers and free inland navigation. Generally referred to as the Salomón–Lozano Treaty, the agreement granted Colombia frontage on the Amazon River in return for ceding Peru territory south of the Putumayo River which Colombia had received from Ecuador in 1916. It also provided for the creation of a mixed commission to mark the boundary and granted the signatories freedom of transit by land as well as the right of navigation on common rivers and their tributaries. The 1922 treaty generated considerable public interest, but its terms were not well understood. Debate in Peru focused on the decision to give Colombia frontage on the Amazon River when a more significant consideration was the extent to which the treaty undermined Ecuadorian claims in the *Oriente*. The territory south of the Putumayo River, ceded by Colombia to Peru, comprised the bulk of the area disputed by Peru and Ecuador. Its acquisition by Peru greatly enhanced the Peruvian Government's position in the region *vis-à-vis* Ecuador. Overnight, Ecuador found itself confronted by an antagonist (Peru) where previously it had an ally (Colombia). From the San Miguel River eastward, Ecuador was now enclosed on the north, east and south by Peruvian territory (see Figure 8.1). In addition to destroying any legal support which the 1916 Colombia–Ecuador Treaty had given

Ecuadorian claims, the 1922 treaty eliminated the possibility of Colombian support, either military or diplomatic, for Ecuador in its dispute with Peru. While few Peruvians acknowledged the importance of this new geographical and political reality, the violent reaction that news of the agreement produced in Ecuador testified to its strategical importance. When the provisions of the 1922 treaty finally became public knowledge in 1925, the Ecuadorian Government protested loudly; and after Colombia ratified the pact later in the year, Quito severed diplomatic relations (Muñoz Vernaza 1928: 90-2 and 101; Ministerio de Relaciones Exteriores, Peru 1936b: vol. I, 251-4).

In early 1913, the Peruvian Government had proposed to Ecuador what later came to be known as the 'mixed formula' because it consisted of both a direct settlement and a limited arbitration. Eventually, talks renewed in 1919 led to the conclusion on June 21, 1924, of a new agreement known as the Ponce-Castro Oyanguren Protocol. It provided for the implementation of the mixed formula as soon as the Tacna and Arica dispute was resolved. With the prior assent of the United States Government, the signatories agreed to convene in Washington to negotiate a definitive boundary; and where they were unable to settle, they agreed to submit the unresolved segments to the arbitral decision of the United States. The 1924 protocol attempted to reconcile Peruvian insistence on a legal arbitration with Ecuadorian insistence on a equitable arbitration through direct negotiations; unfortunately, the agreement was neither clear nor satisfactory. In consequence, the positions of both Peru and Ecuador after 1924 continued to reflect the terms of the Spanish arbitration. Confident in its legal title, Peru emphasized a legal arbitration of the dispute while Ecuador, now certain that its legal arguments would not give it frontage on the Amazon River, insisted on a equitable arbitration in the form of direct negotiations. Initially hailed a diplomatic victory in both countries, the 1924 agreement soon attracted growing criticism in Ecuador where detractors challenged its ambiguous provisions as well as the delay which resulted from subordinating the Ecuador-Peru dispute to a resolution of the Tacna-Arica question (Tudela 1941: 38-43; Ministerio de Relaciones Exteriores, Peru 1936b: vol. I, 278-9; Pérez Concha 1961: vol. II, 9-11 and 61-3).

In late 1933, the Peruvian Government invited Ecuador to open negotiations in Lima in accordance with the terms of the 1924 Ponce-Castro Oyanguran Protocol. In the belief that the United States Government would support its claims, Ecuador reluctantly accepted the Peruvian proposal, and in April 1934 a series of desultory talks opened in the Peruvian capital. Unable to find common ground, the negotiations broke down completely in August 1935 with Ecuador withdrawing its delegation. For the next eighteen months, the two governments argued over the nature of the dispute and the form future proceedings should take. Finally, on July 6, 1936, they agreed to take the dispute to Washington for a *de jure* arbitration during which both states would maintain their existing territorial positions. The Washington Conference lasted two long years, and more than anything else it proved to be a test of patience and an exercise in futility. Both the sessions and the proposals were long, repetitious, boring and unproductive. Nonetheless, they did produce a clear statement of the seemingly irreconcilable positions of Peru and Ecuador (St John 1970: 429–54).

Initially, the Ecuadorian delegation maintained that the central issues were territorial as they involved the ownership of large areas of the *Oriente*. In short, Quito hoped to negotiate the possession of the entire territory north of the Tumbes, Huancabamba and Marañón Rivers. According to the Ecuadorian delegate, the two governments had come to Washington to negotiate a comprehensive direct settlement or a partial settlement to be followed by a limited arbitration by the president of the United States. Later, Ecuador proposed a complete juridical arbitration of the dispute. While this proposal suggested a shift in its attitude towards arbitration, it was largely an attempt to precipitate a Solomon-like judgment by the United States Government. In contrast, the opening statement of the Peruvian delegation emphasized that the dispute was not one of organic sovereignty but rather one of frontiers. According to Peru, the issue at hand was the exact location of the boundary line between the three Peruvian provinces of Tumbes, Jaén and Maynas and adjacent Ecuadorian territories. This was the same position the Peruvian Government had taken in the Spanish arbitration four decades earlier. When Peruvian Foreign Minister Carlos Concha eventually announced the termination of the Washington Conference, he explained that it was impossible for Peru to continue because Ecuador's proposal for total arbitration was outside the spirit and letter of the 1924 protocol, a pact which contemplated only an eventual and partial arbitration by the president of the United States. He added that the only legitimate areas for discussion remained the exact limits separating Tumbes, Jaén and Maynas from adjacent Ecuadorian territory (Ministerio de Relaciones Exteriores, Ecuador 1937: 5–6 and 43; Ministerio de Relaciones Exteriores, Ecuador 1938: 219–76; Ministerio de Relaciones Exteriores, Peru 1938: 10–11, 25–81 and 229–32).

THE BORDER WAR OF 1940-1

From 1940 to 1941, border incidents along the unmarked jungle frontiers increased as both Peru and Ecuador asserted their territorial claims in the disputed region. Ecuadorian incursions produced numerous skirmishes and battles with Peruvian units. This situation of increasing tension on the border was accompanied by an aggressive press campaign in Quito which charged that Peru was preparing for war. As both the political and military situation deteriorated, the governments of Argentina, Brazil and the United States offered their good offices in an effort to contain the conflict. While the Peruvian Government accepted the offer, it was with the understanding that Peru intended to retain Tumbes, Jaén and Maynas. Willing to accept good offices to reduce the possibility of war, Lima rejected an Ecuadorian suggestion that this procedure be employed as the basis to negotiate a final solution. Hostilities opened in early July 1941 in the Zarumilla sector with both sides claiming the other fired the first shot. The conflict spread quickly as Ecuador launched new attacks in the eastern sector along the Tigre and Pastaza Rivers. After intense fighting on several fronts, Peruvian forces blocked the Ecuadorian advance and successfully counter-attacked. Peru's swift and overwhelming defeat of the Ecuadorian Army was the result of the military reorganization the Peruvian armed forces had undergone in the 1930s as well as the vast superiority of forces it achieved in the main theatre north of Tumbes. In contrast, the Ecuadorian Army, which was largely unprepared for war, suffered from a lack of war *matériel* as well as limited civilian support for the war effort. By the end of July, Peru had advanced some forty miles and occupied four hundred square miles of the disputed territory (Ministerio de Guerra, Peru 1961: 71-2; Pérez Concha 1961: vol. III, 9-112) (see Figure 8.1).

With the outbreak of hostilities, the governments of Argentina, Brazil and the United States, later joined by Chile, worked to organize a peaceful settlement. Their efforts were rewarded on October 2, 1941, when representatives of Peru and Ecuador signed an armistice at Talara. Peace negotiations held in Rio de Janeiro in early 1942 produced a Protocol of Peace, Friendship, and Boundaries. Within fifteen days, Peru agreed to withdraw its forces to a designated area after which technical experts would mark the boundary outlined in the protocol. Under the terms of the settlement, the governments of Argentina, Brazil, Chile and the United States agreed to guarantee both the protocol and its execution. On February 26, 1942, the Peruvian Congress unanimously approved the Rio Protocol, and ratifications were exchanged on April 1, 1943. The mixed Ecuador–Peru demarcation commission was installed in Puerto Bolivar on June 1, 1942; but while the border was soon marked in the west, the demarcation of the *Oriente* was never completed. The Rio Protocol was a major diplomatic victory for Peru as it confirmed Peruvian ownership of most of the disputed territory. In Ecuador, the settlement was widely condemned; and subsequent Ecuadorian governments repeatedly asserted that Ecuador had been and remained an Amazonian state (Pérez Concha 1961: vol. III, 112–394; Ministerio de Relaciones Exteriores, Peru 1967: 27–30).

THE RIO DE JANEIRO PROTOCOL AND ITS SEQUELS

In 1960, José María Velasco Ibarra, a three-time president of Ecuador, initiated a critical and destructive campaign for re-election in which he asserted the Rio Protocol could not be executed. Velasco's arguments focused on a geographical flaw in the 1942 agreement. In the Cordillera del Cóndor region, the protocol defined the border as the *divortium aquarum* between the Zamora and Santiago Rivers; however, aerial surveys subsequently placed the Cenepa River where the watershed was originally thought to be. Once the size and location of the Cenepa River were known, the Ecuadorian Government concluded that the execution of the protocol in that sector was impossible. As a result, a section some forty-five miles long was never marked (see Figure 8.1). In 1951, Ecuador's President Galo Plaza had used this discrepancy as a justification for declaring that Ecuador could never accept a final boundary which did not recognize its rights to a sovereign outlet to the Amazon through the Marañón River.

A decade later, the Velasco administration seized on the misunderstanding to declare the entire border in doubt and the execution of the Rio Protocol impossible. In August 1960, after winning a major popular victory in the presidential elections, President Velasco declared the Rio Protocol null and void. One month later, the Ecuadorian foreign minister argued that Peru and Ecuador must return to the terms of the 1829 treaty which had fixed the Amazon River as its natural boundary. At the same time, he repeated allegations that the Rio Protocol was unjust, imposed by force, and impracticable (Zarate Lescano 1960: 61–79; Chirinos Soto 1968: 7–29; St John and Gorman 1982: 188–9).

In October 1976, the Ecuadorian ambassador to the United Nations demanded a renegotiation of the 1942 Rio Protocol on the grounds that Peruvian occupation of the *Oriente* blocked Ecuadorian access to the Amazon River network and thus severely limited its participation in the economic development of the region. At about the same time, the United States Government complicated the dispute by suggesting that the Peruvian position was too radical, thus encouraging Ecuador to think that a compromise solution in its favour might be possible. Mounting tension between Peru and Ecuador finally led to skirmishes in and around Paquisha in the Cordillera del Cóndor region in January 1981 from which Peru emerged triumphant militarily. While the Peruvian Government took decisive action to defend its national patrimony, the terms of the subsequent cease-fire were criticized in Peru for not providing for a demarcation of the boundary, for not referring to the legal principle of respect for international agreements and for not involving the guarantors of the 1942 Rio Protocol. This concern resurfaced in October 1983 when the Ecuadorian Congress again declared the 1942 Protocol null and void and reaffirmed Ecuador's territorial rights in the Amazon Basin. In recent years, diplomatic and commercial relations between Ecuador and Peru have improved markedly; but further progress on issues involving national security appears to await resolution of Ecuador's Amazonian pretensions (Mercado Jarrín 1981: 22–106; Ferrero Costa 1987: 64–5).

CONCLUSION

The boundary dispute between Ecuador and Peru has persisted for well over a century and a half. Over that period, Peru possessed the stronger *de facto* case as it occupied and developed Tumbes and Jaén after 1822 as well as much of Maynas. In addition, Peru clearly developed the superior *de jure* case to the contested territories. This was confirmed by the projected award of the Spanish arbitration in 1910. Accepting this fact, Ecuador thereafter insisted on an equitable solution through arbitration or direct negotiations. Thought to have been resolved in 1942, the question remains today a major issue on the foreign policy agenda of both states. At the same time, the character of the dispute has changed completely in the last half century. Since the conclusion of the Rio Protocol, the case is closed from a legal standpoint. In seeking to void unilaterally a recognized treaty of limits, the Ecuadorian Government is challenging a rule of international law whose overthrow could signal chaos for a region where dozens of such treaties have been negotiated since independence. It is as a political issue that the dispute lives on, since Ecuador remains determined to satisfy what it considers to be its moral rights in the Amazon Basin. To this extent, the Ecuador–Peru boundary dispute will continue as a lively issue in the foreseeable future, disturbing sub-regional relations until a compromise solution is found.

NOTES

1   The available evidence does not support the contention of Maier (1969: 34) that Peru based its legal claim on the *cédula* of 1802 as its claim here was always secondary to title based on the principle of self-determination.
2   Drouet (1936: 31 and 65–9) was wrong to suggest that Peru generally accepted the year 1810 for the commencement of *uti possidetis* as this was never true in the case of the Ecuadorian dispute.
3   Maier (1969: 38) is inaccurate to imply that the 1829 treaty was duly ratified by both signatories as Greater Colombia's ratification was clearly imperfect.

BIBLIOGRAPHY

Arroyo Delgado, A. (1939) *Las negociaciones limítrofes Ecuatoriano-Peruanas en Washington*, Quito: Editorial de 'El Comercio'.
Basadre, J. (1968) *Historia de la República del Perú*, 16 vols, Lima: Editorial Universitaria.
Brierly, J.L. (1963) *The Law of Nations: An Introduction to the International Law of Peace*, Oxford: Oxford University Press.
Burr, R.N. (1965) *By Reason or Force: Chile and the Balancing of Power in South America,1830–1905*, Berkeley and Los Angeles: University of California Press.
Cano, W. (1925) *Historia de los límites del Perú*, Arequipa: Tipografía Quiróz Perea.
Checa Drouet, B. (1936) *La Doctrina Americana del Uti Possidetis de 1810*, Lima: Librería e Imprenta Gil.
Chirinos Soto, E. (1968) *Perú y Ecuador*, Lima: Talleres Gráficos P.L. Villanueva.
Cornejo, M.H. and de Osma, F. (1909) *Memorandum final presentado por los plenipotenciarios del Perú en el litigio de límites con el Ecuador*, Madrid: M.G. Hernández.
Deler, J.-P. (1981) *La Genèse de l'espace équatorien*, Institut Français d'Etudes Andives, Synthèse No. 4, Paris: Editions ADPF.
Eguiguren, L.A. (1941) 'Notes on the international question between Peru and Ecuador: Part I, Maynas', Lima, n.p.
Ferrero Costa, E. (1987) 'Peruvian foreign policy: current trends, constraints and opportunities', *Journal of Inter-American Studies and World Affairs*, 29, 2: 64–5.
Flores, P. (1921) 'History of the boundary dispute between Ecuador and Peru', unpublished Ph.D. thesis, Columbia University.

García Salazar, A. (1928) *Resumen de historia diplomática del Perú, 1820–1884*, Lima: Talleres Gráficos Sanmartí y Cía.

Lescano, J.Z. (1960) *Reseña histórica del problema limítrofe peruano-ecuatoriano*, Lima: Ministerio de Guerra.

Maier, G. (1969) 'The boundary dispute between Ecuador and Peru', *The American Journal of International Law*, 63.

Mercado Jarrín, E. (1981) *El conflicto con Ecuador*, Lima: Ediciones Rikchay.

Ministerio de Guerra, Peru (1961) *Biblioteca Militar del Official no. 31: estudio de la cuestión de limites entre el Perú y el Ecador*.

Ministerio de Relaciones Exteriores, Ecuador (1937) *Las negociaciones ecuatoriano-peruanas en Washington setiembre 1936–julio 1937*, Quito: Imprenta del Ministerio de Gobierno.

—— (1938) *Las negociaciones ecuatoriano-peruanas en Washington agosto 1937–octubre 1938*, Quito: Imprenta del Ministerio de Gobierno.

Ministerio de Relaciones Exteriores, Peru (1890) *Memoria del Relaciones Exteriores del Perú*, Lima: various publishers.

—— (1896) *Memoria del Relaciones Exteriores del Perú*, Lima: various publishers.

—— (1936a) *The Question of the Boundaries Between Peru and Ecuador: A Historical Outline Covering the Period Since 1910*, Baltimore: Reese Press.

—— (1936b) *Tratados, Convenciones y Acuerdos Vigentes entre el Perú y otros Estados*, 2 vols, Lima: Imprenta Torres Aguirre.

—— (1937) 'Resumé of the historical–juridical proceedings of the boundary question between Peru and Ecuador', Washington, DC, n.p.

—— (1938) *Documentos relativos a la conferencia Peru-Ecuatoriana de Washington*, Lima: Talleres Gráficos de la Editorial 'Lumen'.

—— (1967) *Protocolo Peruano-Ecuatoriano de paz, amistad y límites*, Lima: Tipografía Peruana.

Muñoz Vernaza, A. (1928) *Exposición sobre el tratado de límites de 1916 entre el Ecuador y Colombia y análisis jurídico del tratado de límites de 1922 entre Colombia y el Perú*, Quito: Talleres Tipográficos de El Comercio.

Pérez Concha, J. (1961) *Ensayo histórico-crítico de las relaciones diplomáticas del Ecuador con los Estados Limítrofes*, 3 vols, Quito: Editorial Casa de la Cultura Ecuatoriana.

Porras Barrenechea, R. (1942) 'El litigio Perú-ecuatoriano ante los principios jurídicos Americanos', Lima, n.p.

St John, R.B. (1970) 'Peruvian foreign policy, 1919–1939: the delimitation of frontiers', unpublished Ph.D. thesis, University of Denver.

—— and Gorman, S.M. (1982) 'Challenges to Peruvian foreign policy', in Gorman, S.M. (ed.) *Post-Revolutionary Peru: The Politics of Transformation*, Boulder: Westview Press.

San Cristóval, E. (1932) *Páginas internacionales*, Lima: Librería e Imprenta Gil.

Santamaría de Paredes, V. (1910) *A Study of the Question of Boundaries Between the Republics of Peru and Ecuador*, Washington DC: Byron S. Adams.

Soder, J.P. Jr (1970) 'The impact of the Tacna–Arica dispute on the Pan-American Movement', unpublished Ph.D. thesis, Georgetown University.

Tobar Donoso, J. and Tobar, A.L. (1961) *Derecho territorial ecuatoriano*, Quito: Editorial 'La Unión Católica'.

Tudela, F. (1941) *The Controversy Between Peru and Ecuador,* Lima: Imprenta Torres.

Ulloa Cisneros, L. (1911) *Algo de historia. Las cuestiones territoriales con Ecuador y Colombia y la falsedad del protocolo Pedemonte-Mosquera,* Lima: Imprenta La Industria.

Ulloa Sotomayor, A. (1941) *Posición internacional del Perú,* Lima: Imprenta Torres Aguirre.

—— (1942) *Perú y Ecuador: ultima etapa del problema de límites (1941–42),* Lima: Imprenta Torres Aguirre.

Wagner de Reyna, A. (1962) *Los límites del Perú,* Lima: Ediciones del Sol.

—— (1964) *Historia diplomática del Perú, 1900–1945,* 2 vols, Lima: Ediciones Peruanas.

Wood, B. (1978) *Aggression and History: The Case of Ecuador and Peru,* Ann Arbor: University Microfilms International.

Wright, L.A. (1941) 'A study of the conflict between the Republics of Peru and Ecuador', *Geographical Journal,* 98: 253–4.

Zook, D.H. Jr (1964) *Zarumilla-Marañón: The Ecuador–Peru Dispute,* New York: Bookman Associates.

9

TECHNOLOGY, GEOPOLITICS AND FRONTIERS IN BRAZIL

Bertha K. Becker

INTRODUCTION

Assuming that capitalism is a social system whose concepts are rooted in history and that its development addresses major economic, social and political issues, three parameters can be identified for a modern interpretation of space and frontiers in a capitalist society.

First, the changes taking place today are the result of two interrelated processes: the technological revolution and the restructuring of the bases of the accumulation model, characterized by the ever-increasing internationalization of the capitalist world economy. Science and technology, embedded in the social structures of power, have become key elements in the current world situation. A web of managerial flows and technological vectors tends to replace concepts of place, frontier and the State (Castells 1985).

Second, although the logic of modernization is an equalizing factor and the State serves as mediator between economic globalization and national territory, spatial differences tend to increase and states continue to be essentially political units. Any analysis of the relationship between the scientific-technological vector and national territory must consider the particularity of social formations. It is at this level that major tensions develop between local economies and the world system which is in turn mitigated by a transnational structure of monetary–commercial–informational circulation and by the power play of the State. Frontiers present a clearer vision of the tension between the global economic changes and territorial space, embodied by the sovereignty of the State.

Third, the scientific-technological model of territorial development is not the result of the free play of market forces, but rather of political players, decision makers and organizational strategies. The actions of these players have specific characteristics, inherent to the different social formations to which they belong. In the case of Brazil, these characteristics are intimately related to the structure of the national State and to the unique role played in it by the armed forces.

Within this context, the territorial configuration of political changes can no longer be explained by traditional analyses and concepts. Geopolitics become incomprehensible without considering the modern scientific-technological vector, that is, the control of time. The power associated with speed and accelerated communications, which is so closely related to the war machinery, is indeed the very essence of the technological revolution (Virilio 1983). As a result, the concept of frontiers also changes, and is displaced by the space–time concept.

Several questions arise from this context: what do frontiers mean in modern times and what specific features do they have in Brazil? Two hypotheses are proposed. First of all, frontiers are defined by a specific space at a specific time, where the cohesion of structures may vary widely from one region to another. This is why changes occur so rapidly in space–time. Second, in Brazil, frontiers have constituted the space–time setting in which the State has chosen to promote conservative modernization over the past twenty years.

These hypotheses are discussed in four sections. In the first section, the meaning of frontiers is explained in a context of conservative modernization. In the second and third sections, the concept of conservative modernization is illustrated by its two main expressions on the frontier: scientific-technological and techno(eco)logical. The last section discusses frontiers from the perspective of the crisis being faced by the State in the context of the new global situation.

THE MEANING OF THE FRONTIER IN CONSERVATIVE MODERNIZATION

As a result of its historical development, Brazil, like many other Latin American nations, experienced a late and authoritarian type of capitalism. It is considered late because economic growth was not accompanied by the specific production forces of capitalism; authoritarian in that the State, and a powerful bureaucracy where the armed forces play a key role, politically dominates relations within society. Maintaining a delicate balance between world market forces and the interests of the dominant groups conferred increasing powers on the State, which assumed a decisive role in fomenting industrialization.

Consequently, the building of the State preceded the building of the nation. Conservative modernization is, thus, the Latin American approach for achieving modernization, the approach by which the State negotiates privileges with private groups, deciding whether to include them or not in public benefits, in return for their support for 'top-down' modernization.

Brazil is a paradigm of this approach. It exhibits the most complex late capitalism in the hemisphere, as its modernization policies illustrate well. State policies are invested with a specific territoriality which is at the very root of the Brazilian social formation and the study of such policies thus has a geopolitical dimension. In such a context, the frontier becomes a key expression of Brazil's particular brand of late authoritarian capitalism.

Contrary to Turner's concept that views a vast settlement frontier as the key element in the building of American democracy, in Brazil the frontier is historically associated with authoritarianism (Velho 1979), although it has been defined differently at distinct times.

Political frontiers were always at the fore of the spatial articulation of production. Control over an extensive territory, which remained undivided at the time of independence, was due to the interests of a pro-slavery oligarchy which made up the Brazilian Empire. At the same time, the expansion of the economic frontier contributed to pacts among the oligarchies, increasing agricultural exporting activities through the incorporation of new land and the creation of large landed estates without affecting the pre-existing land structure (Becker 1982). For the pro-development sector in Brazilian society, the intervention of the modern State in industrialization, space and frontiers became key requisites for accumulation and a symbolic recourse for legitimacy. In the 1930s, the 'march to the west' discourse contributed not only to the centralization of government power under the authoritarian regime of the 'New State', but also to bolstering the spontaneous expansion of the agricultural frontier in order to supply the two main industrial centres, Rio de Janeiro and Sao Paulo, with cheap labour. This process culminated with the national development movement of the 1950s, and the construction of the new capital, Brasilia (1960), a symbol of State power, established in the interior highlands of Brazil with a view to extending and legitimizing its power throughout the national territory.

The movement of frontiers contributed, therefore, to late capitalism and authoritarianism. On the one hand, the expansive use of space was a strategy to optimize the work/product ratio, thus partially compensating for the absence of technical capitalization. On the other hand, it supported oligarchic pacts and, recently, in terms of the domestic market, it served to centralize government power and contributed to the construction of the State itself. Nevertheless, in the past two decades the relationship between the State and frontiers has changed. A new type of authoritarianism has emerged and frontiers have taken on a new meaning.

The State of the 1930s and of the 1950s became both a promoter of and an actor in the industrialization process. By the mid-1960s, it took upon itself the initiative to establish a national modernization project, thus anticipating the social changes this would produce. In other words, the State itself began to generate new space and frontiers through the deliberate policies of a socially exclusive military regime.

In 1964, the armed forces took over the State apparatus and stayed in power for twenty-one years. In alliance with the industrial and financial middle class and legitimized by the bourgeoisie and the oppressed working class, the military regime inaugurated a period of new-style authoritarianism. In the new authoritarianism that emerged in Latin American societies undergoing modernization in the 1960s, particularly in the countries of the Southern Cone, the decisive factor was the militarization of the State. The military no longer ruled at the individual level, but rather as an institution and with a bureaucratic approach in formulating policies. The military took power with the aim of restructuring society, and the State intervened in grassroots movements, keeping the uninterrupted progress of development subject to the modern military doctrine of national security (Cardoso 1979).

External conditions and the dynamics of internal class struggles fostered the emergence of the new type of authoritarianism in Brazil.[1] Organizational conditions and the timing of policies were essential to this model. The political role of the armed forces in Brazil can be explained, in part, by its ability to organize, compared to the ideological weakness and political indecision of the leading opposition groups and to the weakness of grassroots organizations (Carvalho 1985). The military forces took over the State because they were organized organically and because they had a project: a national, geopolitical project geared towards modernization.

The premises of the geopolitical project were not determined by the geography of the country nor by the acquisition of territory, as in traditional geopolitics. The new project was intended to control the modern scientific-technological vector; that is, to control not only space, but above all to control time. In this context, control is understood to be a requisite for establishing a new paradigm to insert the country into the new world political order of the post-war era, and for the rapid modernization of society and of the national territory. It is also a requisite for strengthening and expanding the leading role of the State, as the only actor capable of fostering rapid modernization through efficient planning (Becker 1988a).

It is not a question of making the military omniscient. Born at the height of the liberal post-war regime, the modernization project was hardly the fruit of the armed forces, nor was it the result of rational and intelligent progress; rather it arose from isolated initiatives and decisions made in response to existing conditions that led to a project managed by the military. Nevertheless, two aspects of the military's participation should be noted: a) the recognition of science and technology as the fundamental basis of national security – i.e., to achieve economic growth and international importance; and b) isolated initiatives, introduced in the late-1940s, developed into a complex scientific-technological system in the 1970s (Becker and Egler 1989).

Thus, rapid modernization resulted from a combined movement: on the one hand, foreign investments and abundant loans 'pushed' by large international banks during the 1970s[2] and, on the other hand, the geopolitical modernization project that gave the State the technical capabilities to deal with space–time on a large scale. The concrete expression and foundation of this movement was the 'triad' – the association of State capital, foreign capital and private national capital.

But how could significant changes be achieved without disturbing the hierarchical social order? That is, how could conservative, rapid modernization be promoted? The answer was through production and the management of space and, particularly, by opening frontiers.

The State became the mediator between the new world economy and Brazil's social formation. It introduced rapid changes through flows of international capital, goods and information, although the established structures tended to delay the rate of modernization (Becker and Egler 1990). The State treated space as an integral and fundamental part of the technological base of a large oligarchic enterprise in an attempt to provide it with the operational and functional capabilities needed to guarantee not only handsome profits for the parties involved in the process, but also the integration of the national territory by opening up favoured areas for investment for the world economy (Egler 1983).

Within this context, the politization of the spatial structure was taken to the extreme. In order to coordinate multiple times and spaces that correspond to different interests, a complex type of territory management developed, involving not only management in economic territories, but also in those related to power. The management of territory, understood as the strategic, scientific and technological exercise of power to control space–time, was an essential tool of conservative modernization in that it consolidated a major world metropolis, Sao Paulo, home of the 'triad', as a link to the world economy. Furthermore, it guaranteed the necessary agricultural and commercial opportunities to sustain the 'establishment' by encouraging the expansion and colonization of frontiers to further spatial change without threatening established interests.

The articulation of the metropolis, territories and frontiers was achieved by the State imposing a vast network politically and technically controlled by government programmes and projects. As an integral part of a programmed network, these government projects were designed to address the interests of the components of the 'triad'. These were principally a) the extension of all types of networks – road, urban, communications, information, institutional, banking, etc., and b) the creation of territories superimposed on the official political–administrative divisions, supervised by federal institutions, through which investments could be channelled (Becker 1988b).

National integration is seen here as a territorial ideology and the word 'frontier' takes on new meaning. It is not limited to tracts of pioneer settlements in 'free' open lands; nor is it simply a type of peripheral area. It takes on a symbolic dimension at the national level. By incorporating the pioneer utopia and manipulating the perception of space, the State adopts and disseminates, through the media, the term 'frontier' to hallow the expansion of society and territorial integration. It raises expectations and exacerbates social tensions, redirecting them towards the 'empty spaces' of the interior, giving the frontier the image of a space of new opportunities for the nation, i.e. of vertical mobility which it was impossible to achieve in Brazilian urban society (Becker 1986 and 1988a). Thus, the frontier became a component of the material and ideological patrimony that determines the relationship of society with its territory as a concept closely linked to the founding myth of a given society (Aubertin and Lena 1986).

In practice, however, the frontier, as a privileged battlefield of the State, ensures the expansion of its power; that is, it becomes an essential element in the production of space. Thus, frontier can be defined as a not fully structured economic, social and political space and, therefore, the possible generator of new realities (Becker 1986 and 1988a). Moreover, as a space of accelerated processes, the frontier is where the State can more rapidly attain its objectives of conservative modernization. The geopolitics of the State in Brazil built not one, but many frontiers that provided perspectives for economic growth, a solution to social tensions and the full control of time and space. Among these perspectives, mention must be made of the scientific-technological frontier, associated with the weapons industry, implanted in the 'core area' of the south. While creating a new model for integration in the world order, this technological frontier was not sufficient to achieve the integration of Brazil's vast territory. These integrationist and modernist policies reached their maximum expression with the opening of the Amazon frontier.

THE SCIENTIFIC-TECHNOLOGICAL FRONTIER[3]

According to Hirschman (1986), the industrial policy of 'import anticipation' refers to the decision to establish a 'short-circuit' in the import-substitution process by stimulating the domestic production of key capital goods. This was the result of the armed forces' desire to control the scientific-technological vector in four strategic sectors, aerospace, arms, nuclear energy and computing, by opening up the scientific-technological frontier.

The aerospace industry resulted from the creation of the Instituto Técnico da Aeronáutica (ITA) in 1946, which became the Centro Tecnológico da Aeronáutica (CTA) in 1971, the explicit purpose of which was to integrate teaching, research and industry in a long-term coordinated action. The CTA became one of the most important institutions in the country in terms of human-resource training, and the projects it developed gave rise to an aeronautical industry, with the creation of EMBRAER (Brazilian Aeronautical Enterprise) in 1969. It became a focal point for the establishment of other technological centres, such as the National Space Research Institute (INPE), and for a large number of both large and small multinational and national companies. In 1971, INPE was successfully incorporated into the Aerospace Technological Center.

As a result of the Vietnam War, the United States curtailed its exports of weapons and its credit for arms. This, together with fear of domestic threats, encouraged the Brazilian military to stop importing weapons and to produce them, adapting them better to the conditions of the country. Public funds and protectionist measures were geared to stimulating this sector. Scientific and technological activities, linked to the manufacture of weapons in the public and private sectors, were also encouraged. The army followed its own path towards modernization.

The ill-equipped and spatially inarticulate armed forces tried to streamline their modernization project through the establishment of IMBEL in 1975, a military state corporation created to produce different types of equipment for ground forces in association with private companies. This stimulated the production of explosives, armoured vehicles and tanks. Consequently, in less than a decade Brazil became one of the world's ten largest exporters of weapons, catering chiefly to Iraq.

Nuclear research in Brazil dates back to the creation in 1951 of the National Research Council (CNPQ), which was designed by an admiral with the purpose of developing a nuclear programme independent of the United States. A Brazil–Germany agreement was signed in 1975, under the responsibility of a state-run company, NUCLEBRAS. Given the failure of this initiative due to technical problems, the Brazilian Navy secretly began to develop a 'parallel' nuclear programme in 1979. The Navy also played an important role in the national policy on the computer industry. The first steps were taken in 1971 and, in 1979, the National Computer Science Centre was created, aimed at establishing a policy for the sector. The domestic market was temporarily reserved for domestic firms. The microcomputer industry grew rapidly and developed the capacity for innovations. It became an important source of employment as well as a source of conflict with the United States, which is only now being resolved as a result of the new world situation.

The Second National Development Plan (1975–9), established in the midst of the first petroleum shock, integrated an 'import anticipation' policy for technology. Compared to most countries, which adopted recessionist policies, the government of Brazil set the economy on a 'forced march', implementing large projects with external funding (Castro and Souza 1985). Efforts related to science and technology were, for the most part, aimed at state and military research and development (R. & D.) centres. New ones were created and existing ones were modernized in order to control strategic sectors such as aerospace, mineral mining, nuclear energy, oceanography, petrochemicals, electronics, weaponry and telecommunications. These were closely linked to research developed in certain private companies and universities (in Sao Paulo, Camoinas, Sao Carlos and the city of Sao Paulo and in Rio de Janeiro the Universidade Federal e Católica).

The scientific-technological frontier, the 'locus' of the geopolitical modernization project, was initially implemented in the Paraiba do Sul Valley, historically the main link between the large cities of Rio de Janeiro and Sao Paulo, where the aim was to build a military-industrial complex. CTA and an important unit of IMBEL were located in the valley, as well as other important private industries that produce missiles (Avibras) and tanks (Engesa). Today, 80 per cent of the weapons industries are concentrated in this valley.

The decision to locate the new project in the Paraiba do Sul Valley is due to its strategic location and to its favourable conditions in terms of the availability of land and an appropriate technical environment. The valley serves as one of the principal routes to the interior plateau of Brazil and to the metropolitan corridor which is a vital national distribution point. It is also close to two large information command centres, as well as the industrial core of Sao Paulo and the military decision-making centres of Rio de Janeiro. Technical experts, trained in old military institutions and in the first large iron and steel industry of Volta Redonda, founded in 1942, contributed to making this an essential location. Also involved were the sons of the local elite, heirs to a glorious era of coffee growing, who participated in implementing the project.

Consequently, the territorial division of labour was redefined in the valley and its surrounding areas, based on science and technology. Mention should be made of an aerospace pole, a metalmechanics centre, a chemical and explosives industry zone and an area designated for electric and electronic production. Also located in the vicinity of Sao Paulo, but in the opposite direction – to the west, was the Aramar Navy Research Centre. The main cities of Sao Paulo and Rio de Janeiro furnished sophisticated inputs for the industries in the valley and, together with its R. & D. centres, formed the embryo of the scientific-technological frontier which, today, extends into the interior of the state of Sao Paulo.

THE TECHNO(ECO)LOGICAL FRONTIER: THE AMAZON RIVER REGION

Today, the Amazon River Region is the basic expression of Brazil's frontier. National integration placed high priority on inhabiting the vast, low-population-density forest area, which corresponds to more than half of the territory, with a view to promoting an internal and external geopolitical balance. In this way, it was possible to resolve simultaneously problems linked to social tensions in the periphery, the growth of the core area and the leadership in South America (Becker 1982). In fact, the territorial occupation of this area, this time on a gigantic scale and in such a rapid manner, was essential to maintaining the authoritarian modernization model. The rationale behind this territorial strategy was that it would make it possible to a) eliminate revolutionary hot spots, b) create an alternative for investments, principally in land, for ranchers and businessmen, c) avoid agrarian reform, which made it possible to re-establish large properties, to transfer small-scale food production to the interior of the country and to allow emigration from socially tense areas, particularly in the north east and in the large cities, d) give substance to the ideology of territorial nationalization through the occupation of 'empty spaces', and e) involve Brazil in the South American Amazon River Region, in terms of policies

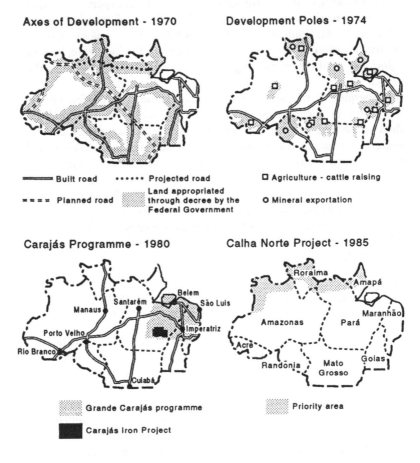

Figure 9.1 Territorial occupation policies in the Amazon

as well as in terms of exploiting resources and exporting manufactured goods.

Due to a lack of social organizations capable of showing resistance, the federal government directly assumed the management of regional modernization, rapidly implementing a large-scale programmed network. In this way, the former central-western and north-eastern regions and all the northern region became one large national frontier.

The components of the programmed network are most transparent in the Amazon River Region in the form of a) large networks for spatial integration – highways, urban, telecommunications – cut through the dense jungle, b) new territories of the federal government replacing state territories, created by decrees under which central power has absolute jurisdiction and the right to property, c) subsidies for capital flows favouring the private appropriation of land by agricultural and mining companies, and d) strong incentives for migrants to occupy the territory for the formation of the labour force. This made it possible to occupy the lands at the frontier in a movement that stopped at the edge of the jungle.

Focal points of modernization are typified by the Manaus Free Zone and, since 1980, by the creation of large mineral mining projects which served as company towns, centres of production and the management of joint ventures or independent corporations. During the 1980s, large projects characterized the State's strategy in the region. They reflect the economic crisis that followed the second petroleum shock (1979) and, with the rapid increase in interest rates on international markets, attempts were made to maintain economic growth through exports. But these efforts are also an expression of the 'forced march' and of the attempt to expand and transnationalize state companies.

The most outstanding example of the new strategy is the Grande Carajás Programme (PGC), which involved the creation of a new territory of ninety-million hectares (or 900,000 sq. km), almost 10 per cent of the national territory, through subsidies and the implementation of a hydroelectric network and a railway. The most important company in the territory is the Vale do Rio Doce state company (CVRD), the world's largest iron-exporting company, which under the PGC diversified its activities and increased its share in the world market. This company is the sole administrator of the Carajás Iron Project (PFC) which includes a two-million hectare territory, a mine, a railroad and a port. It also holds the major share in other joint-venture projects.

The radical implementation of the 'programmed network' resulted in generalized conflicts between indigenous groups, miners, businessmen and corporations that converted the frontier into the setting for global–

local tensions, with the State acting as referee. The size of the two large projects paralleled a change in the scale of conflicts which were no longer just land disputes: individual prospectors vied with companies to preserve areas of manual labour; the 'povos da floresta' (jungle people), who, displaced from their territory, formed the *Uniao dos Povos de Floresta* (1989), demanded the establishment of indigenous lands and *reservas extrativistas* (federally protected areas which granted the right of usufruct to the settlers).

The PFC, managed by the CVRD, is a good example of a company involved in these conflicts. The power of the company is obvious in its control of the immense territory and the mineral reserves found there, evidenced by the presence of a fortress and security belt maintained by the company. Inside this territory is the Serra Pelada, where more than eighty thousand prospectors excavate the land manually to extract gold. A social and technological war erupted between the company and the miners, which pressurized the federal government to make concessions, extending the duration of manual labour in the CVRD territory.

This political deadlock was evident on the border with Bolivia where the rubber tappers of Acre, led by Chico Mendes (now brutally murdered), have occupied the rubber plantations to prevent deforestation. On the border with Venezuela, the Yanomamis Indians have been submitted to all kinds of pressure in their territory, which is rich in gold, uranium and precious stones; mineral explorations by prospectors and companies, transnational religious missions and contraband traffic. In addition, an immense military project – the Calha Norte Project – aimed at mitigating these conflicts and consolidating this border region, covers 14 per cent of the national territory (1,221,000 sq. km) extending 900 km along the political border with Venezuela.

Rapid and intense action on such a scale made the Amazon frontier into an effective arena for State action. Here some of the fiercest and most publicized conflicts took place, associated with the financial and political crisis of the State in the 1980s. It also led to conflicts over sovereignty on a continental scale – concerning no less than 12,000 km of boundaries – involving every South American state except for Chile.

Disorganized settlement movements exacerbated tensions along political boundaries. The opening up of the settlement frontier also fostered illegal activities, including contraband in gold, drugs, wood, animal pelts, electronic products, etc. Migratory movements increased across the political boundaries of each country, flowing from one to the other, depending on the opportunities available in each. Examples of this situation are a) the flow of the labour force to French Guiana,

b) contraband, based on the resale of fuel and the occupation of Venezuelan soil by Brazilian miners, c) drug traffic on the border with Colombia and Bolivia and d) the exploitation of rubber by Brazilian labourers who cross over into Bolivia.

The paradoxes of regional settlement policies and of the national/ transnational economic growth pattern are most evident in the world ecological debates on the Amazon (see Figure 9.2). Ecological debates involve not only the legitimate interests of Indians, rubber tappers and local environmentalists, but also broader technological and geopolitical interests and a growing world environmental movement. The disproportion between these two scales of interest, when faced with the real problem of environmental degradation, forces conflicting interests of national and international players into unusual coalitions. On the one hand, energy development projects gave continuity to the geopolitical policy of frontier expansion, but within a context of crisis in the State, whose austerity programmes, while aimed at quantitative growth, necessarily became selective. These policies were geared at expanding the hydroelectric network and developing iron, steel and charcoal projects, such as those associated with the iron ore extraction venture of the Rio Dorado State Enterprise (CVRD). A biotechnological front also developed, linked to a new technical-scientific paradigm of the capitalist market economy, based on qualitative growth and sustainable use of the vast biomass resources of the Amazon (Becker 1990).

As both a national and world frontier, the Amazon Region takes on a key position as a vast reserve of biological capital in a context of crisis and reorientation of the capitalist system, thus becoming a techno(eco)logical frontier. Environmental protection ceases to be a cause only for Indians, miners and national and international environmentalists. Since the region contains the largest genetic bank in the world, large corporations that dominate the genetic engineering sector clamour for its preservation, interested in delineating forest reserves as 'experimental paradises' for scientific-technological development. The Amazon is also of interest for large firms, as a source of sophisticated products for specific market sectors that seek to move closer to nature. Consequently, new technologies transform the environment and nature into market commodities.

The Amazon frontier is, thus, a theatre where national and international interests converge and where new models of industrialism/ ecodevelopmentism are being experimented. However, the Amazon Region is not Antarctica, which was divided up by the large world powers, it is the fundamental patrimony of Brazil.

Figure 9.2 The South American Amazon Basin

THE CRISIS OF THE STATE AND THE OUTLOOK OF FRONTIERS IN BRAZIL

As a result of the combined action of world capitalism and a national geopolitical project for modernization, Brazil changed its position in the world economy. It ceased to be a peripheral nation and became a semi-peripheral nation, which is an ambivalent, intermediate category in which Brazil becomes the exploiter as well as the exploited, with econ-

146

omically and technically advanced sectors coexisting with traditional and large unequal social sectors. In this context, the State plays a crucial role, not only in trying to distort the world market to the benefit of the dominant groups, but also as a fundamental tool to extract taxes, subsidies and to bring about economic growth (Wallerstein 1979).

The use of territory as an instrument, and particularly the opening up of frontiers, was the basic condition for conservative modernization. Nevertheless, maintaining the 'triad' led the State to extend its regulatory and entrepreneurial function beyond its ability to generate public funds, thus weakening its capacity to maintain territories, open frontiers and negotiate with multinational capital. In other words, the State extended its frontiers beyond its ability fully to control them. Thus, the frontiers contributed to the State's vulnerability as well as to the consolidation and growth of the State.

At the same time, the restructuring of the world economy, inspired by the theses of neo-liberalism based on new production and management technologies and geared to the establishment of supranational spaces, strongly affected the semi-peripheral areas of the world economy, especially those with the particular brand of Brazilian authoritarianism. In Brazil the State funded most of its conservative modernization policies on international credit, thus becoming the world's largest debtor, struggling with a major portion of the external debt and with an enormous public deficit. Thus, the economic crisis took on the character of a profound political crisis. The geopolitical project led by the military was discontinued, in a process accompanied by the transition to a civil regime, which was completed in 1985.

Within this context, frontiers, as State frontiers, were particularly affected. In the scientific-technological frontier, the war industry entered into a profound crisis, reducing its supply to the international market. War materials were produced by half a dozen companies, but exported by only two of them. While in 1987 this industry produced between $570 million and $1.2 million and was the second most important source of foreign currency, with trade fronts in thirty-five countries, particularly in Arab countries, in 1989 production dropped to $157 million. EMBRAER, which exported more medium-sized airplanes to the world civilian market than Brasilia, Bandeirantes or AMX, was affected by the scarcity of financial support from the government and by protectionist measures imposed by the United States, its principal civilian market, in retaliation for Brazil's closed market policy on computers. Private companies such as AVIBRAS, the largest producer of missiles, and ENGESA, the largest producer of armoured vehicles in the West, suffered from the end of the Iraq–Iran War, delay in payments from international clients and from their heavy debts, entering into bankruptcy in early 1990.

Because of the controlled production, efforts to transfer technology were limited. But progress made in R. & D. institutions is, in part, irreversible, as is their influence on numerous industries that had the capacity to convert their production to the civilian sector.

With regard to the Amazon Region, the multiplicity of conflicts on different scales reflect the crisis on the techno(eco)logical frontier. Its potential in terms of biodiversity gave it a strategic value as a primary source for science and technology and for environmentalist ideologies throughout the world. Brazilian society was faced with the challenge of establishing appropriate negotiations to ensure its participation in the new, techno(eco)logical frontier (i.e. in research that could promote the use of this priceless inheritance, with social priority and without endangering the environment).

In March 1990, the new, democratically elected government took office, with a liberal platform that clearly favoured a reduction in the role of the State, in favour of private initiatives geared to a new level of modernity, based on efficiency and competitiveness. But government practice is authoritarian. In the name of liberalism, government intervention was higher than ever in Brazil. What is the outlook of the frontier within this context? Without a doubt, new frontiers will be opened, although the types of these frontiers are not yet clear. Significantly, the military has withdrawn from the political scene, but it remains on the frontiers. The Brazilian Nuclear Programme and the Nuclear Submarine Programme continue as planned and the Aramar Naval Experimental Centre in Sao Paulo was able to enrich uranium to 20 per cent. Also, in the Amazon Region, the armed forces retained their presence in three projects. Two of these projects deal with the political frontier – the Calha Norte and the *Programa de Desenvolvimento da Faixa de Fronteira da Amazonia Ocidental* (PROFFAO) – and the third, *Programa Nossa Natureza* (PNN), is a State-run counteroffensive to environmental pressures, coordinated by SADEM, an agency that replaced the National Security Council.

It is still unclear how far the State can be influential in opening frontiers in Brazil, and what forms they will take. The new dialectics between the space of flows and the space of places indicate that the space/time concept of frontiers goes beyond the limits of traditional geopolitical analyses. Any attempt to achieve the democratic manage-ment of the territory is faced with a challenge: the creation of the necessary conditions so that civil society is not simply the object of the modernization process, by redesigning institutional channels of participation, through a network of linkages between segments of civil society that go beyond municipal, state and even national limits.

NOTES

1 Although the need to 'deepen' industrialization was used to explain the appearance of the new authoritarianism in Brazil, which had already made progress in import substitution, other factors must be taken into consideration. The cold war, the Cuban Revolution, pressure from North America, together with the demands of the people and proposals for grassroots reforms, created the impression of a serious threat among the leading groups and in the military.

2 The world recession that resulted from the oil shocks of the 1970s did not affect the large banks which, faced with the problem of where to invest their petrodollars in a recessive economy, found an alternative in financing Latin American countries, among others.

3 This section is based on studies made by Becker and Egler in 1989 and 1990.

BIBLIOGRAPHY

Aubertin, C. and Lena, P. (1986) 'Présentation', *Cahiers des Sciences Humaines*, 22, 3–4, Paris: ORSTOM.

Becker, B.K. (1982) *Geopolitica da Amazonia*, Rio de Janeiro: Zahar.

—— (1986) 'Signification actuelle de la frontière: une interprétation géopolitique à partir du cas de l'Amazonie Brésilienne', *Cahiers des Sciences Humaines*, 22: 3–4, Paris: ORSTOM.

—— (1988a) 'Nation-building in a "newly industrialized country": reflections on the Brazilian Amazonia case', in Johnston, R.J., Knight, D. and Kofman, E. (eds) *Nationalism: Self-Determination and Political Geography*, London: Croom Helm.

—— (1988b) 'A geografia e o resgate da geopolítica', *Revista Brasileira de Geografia*, 50, 2.

—— (1989) 'Gestion du territoire et territorialité en Amazonie Brésilienne: entreprise d'etat et "garimpeiros" à Carajas', *L'Espace Géographique*, 3: 209–17.

—— (1990) *Amazonia*, Sao Paulo: Atica.

Becker, B.K. and Egler, C.A.G. (1989) 'O embriao do projeto geopolítico da modernidade no Brasil', *Texto 4*, Rio de Janeiro: LAGET.

—— (1990) *Brazil: A New Regional Power in the World Economy*, Cambridge: Cambridge University Press.

Cardoso, F.H. (1979) 'On the character of authoritarian regimes in Latin America', in Collior, D. (ed.) *The New Authoritarianism in Latin America*, New Jersey: Princeton University Press.

Carvalho, J.M. (1985), 'As forças armadas na primeira república: o poder estabilizador', in Fausto, B. (ed.) *História geral da civilização Brasileira*, vol. III, Sao Paulo: Difel.

Castells, M. (1985) 'High technology, economic restructuring and the urban–regional process in the United States', in Castells, M. (ed.) *High Technology, Space and Society*, Urban Affairs Annual Reviews, 28, 11: 40.

Castro, A.B. and Souza, F.E.P. (1985) *A economía brasileira em 'marcha forcada'*, Rio de Janeiro: Paz e Terra.

Egler, C.A.G. (1983) 'Dinámica Territorial recente da industria no Brazil: 1970–1980' in Becker, B.K., Egler, C.G.E., Miranda, M.P. and Bartholo, R.S. (eds) *Tecnologia e gestao do territorio*, Rio de Janeiro: UFRJ.

Hirschman, A.O. (1986) 'The political economy of Latin American development: seven exercises in retrospection', paper given at XIII International Congress of the Latin American Studies Association, Boston.

Velho, O. (1979) *Capitalismo autoritario e campesinato*, Sao Paulo: Difel.

Virilio, P. (1983) *Pure War*, New York: Semiotext(e).

Wallerstein, I. (1979) *The Capitalist World Economy*, Cambridge: Cambridge University Press.

10

THE ARGENTINA–CHILE FRONTIER AS SOCIAL SPACE

A case study of the transAndean economy of Neuquén

Susana Bandieri

INTRODUCTION

The purpose of this chapter is to document the lasting nature of economic, social and cultural relations between the province of Neuquén, in north-western Patagonia, and central Chile. More than a mere natural obstacle or common political boundary, the Andean Range constituted for centuries the axis of an integrated social space. Indeed, since times remote, the territory of Neuquén has functioned as an integral part of Chile's urban regional economy, serving as a hinterland for that country's principal Pacific Ocean ports.

The territorial integration of north-western Patagonia into the Argentinian sphere of influence dates back to the late nineteenth century. The advance of the military forces on the indigenous communities of Patagonia during the punitive campaigns between 1879 and 1885 definitively incorporated the region into the Argentinian territorial system. An immediate consequence of these campaigns was the imposition of a territorial organization in accordance with a new scheme of economic and political domination. New administrative limits were defined with the foundation of towns, and a capital was established as the centre of political authority within this border region. An essential corollary of these integration policies was the imposition of the international boundary separating Argentina from Chile across the rugged Andes. However, the presence of the boundary did not displace cattle raising as the dominant activity. The structural characteristics of this activity allowed for the survival of the old forms of social and spatial organizations. The web of indigenous trading routes across the Andes, derived from the traditional use of resources and social space,

151

did not disappear with the occupation of the space by whites and its subsequent incorporation into capitalist modes of production.

The traditional way of life of this border region as a socially integrated space began to deteriorate during the 1930s when both Chile and Argentina, at different times and for different reasons inherent in their own historical circumstances and international considerations, consolidated their territorial system by enforcing the political boundary. In the 1940s, the situation took on a more permanent nature when the industrialization of the national economies implied greater customs barriers for the region. The breakdown of modes of commerce brought about a general crisis in the dominant cattle-raising activity which seriously affected the different social communities related to it, especially the small producers occupying public lands, who saw their ability to make a profit through direct exchange in the border areas dwindle. These traditional producers became increasingly dependent on successive stages of commercial middlemen in order to have access to the Argentine national market. All this has led to their marginalization and to a massive rural exodus from the interior of Neuquén towards its urban centres.

THE INDIGENOUS COMMUNITIES AND THE SOCIAL ORGANIZATION OF SPACE

The pre-Hispanic tradition of Andean transhumance gave Neuquén characteristic social features already noticeable in the indigenous groups that originally peopled the region. While it is difficult to determine which ethnic groups exercised real dominance over the region at the time of earliest contact, one can mention the process of 'Araucanization' which gradually brought about a notable cultural symbiosis among the tribes on both slopes of the Andes. Prior to this process, Pehuenches, Puelches and Poyas were described by Spanish chroniclers as possessing physical and cultural features distinct from the Araucanians on the western slopes of the Andes, in what is today Chile.

These local ethnic groups suffered a real impact from the physical and cultural penetration of the Araucanians (the Mapuches of Chile). The first incursion was possibly brought about by pressure from Inca groups as far away as the south of Chile, or as a result of the exploitive Indian labour system imposed by the Spanish. Later, these changes in the interior borders of Chile led to the domination by newcomers of the original ethnic groups in the area and to the emergence of hybrid culture groups through transculturation. The second cultural incursion

152

compounded the effects of the first by consolidating pre-existing trade through barter and exchange networks between the different ethnic groups on either side of the Andes Range in Neuquén.

This twofold penetration, physical and cultural, was to end in a true process of symbiosis, called 'Araucanization', which would increase, especially at the beginning of the seventeenth century.[1] The impact of the Araucanian culture became noticeable in the primitive hunting and gathering settlements of Neuquén. Concomitantly, the Mapuche groups of Chile – farmers and potters – modified their own economic organization upon moving to Neuquén, dedicating themselves primarily to hunting and cattle trading, at first informally and later, as the concept of private property took hold in the Argentinian pampa, through cattle rustling and contraband.[2] We can see clearly how the topography of the land conditioned the economic organization of the indigenous population: the continual movement of cattle populations to Chile across the Andean Range became the basic element of the socio-economic organization of these people.

The Chilean landowners needed cattle for their own consumption and for export to other Pacific towns, and the Indians of Neuquén acted as excellent middlemen. The herds, built up by rustling, were led from the humid pampa along trails that crossed the country and entered Chile through a hundred easy passes in Neuquén territory.

According to General Manuel J. Olascoaga, a member of Roca's expeditionary forces and future first Governor of the Neuquén territory:

> the conniving in which our Indians of the mountains and of the pampa indulge and always have indulged with Chilean business-men and even with the Government of Chile [leads them to] come to Buenos Aires, Santa Fé, Cordoba and San Luis, returning with great herds along the far reaches to the South of the Salado, Rio Grande and Neuquén and enter Chile through Villa Rica where the mountain range offers an opening of close to a league wide and which is where the road most used by Indians of the pampa goes, the best road between Buenos Aires and Chile. I have a friend in Valdivia who assures me that along that road, some winters, up to two thousand Indians have crossed.
>
> (Olascoaga 1974: 37–9)

He comments on the magnitude of the Indian movement:

> we were surprised to see roads everywhere which, by the innumerable trails that make them up and the animal debris that

littered them, provide evidence of a continuous and ancient traffic ... the earth trampled and hard, deep ruts two feet apart, intertwined and parallel over an area two miles wide ... all indicating a constant traffic, dating back centuries, of millions of men and animals. These great trails are not the tracks of small nomad tribes who have crossed four or six times a year on their raids. They are great highways between great commercial centers; they are true arteries of communication which channel the life, wealth and progress from one people to another.

(Olascoaga 1974: 165–6)

The decision to integrate Patagonian lands with the national territory by means of a punitive expedition, the so-called Desert Campaign of General Julio A. Roca in 1879,[3] is generally attributed to two main objectives. On the one hand, the aim was to end the rustling activities of the Indians and, on the other, to incorporate new lands for cattle production for foreign markets. One must recall the position of Argentina during the nineteenth century as a dependent capitalist country, producer of raw materials and food. Nevertheless, it is believed that especially in the case of the Neuquén territory, the primary objective of these military campaigns can be more specifically stated. It was a matter of ending decisively the threat represented by the existence of the integrated transboundary socioeconomic space that we have been describing, which directly affected the interests of the landholding bourgeoisie, the most powerful social class in Argentina at the time. The Argentinian bourgeoisie owned the country's largest and most productive landed estates, the cattle from which ended up in Chile thanks to the efficient intermediation of the Indian rustlers, and saw itself as representing the interests of the nation. In one of its protagonists' own words it was necessary to stop the 'bleeding of the wound of the Andes, which is the wound of the Republic' (Olascoaga 1974: 40).

General Roca, leader of the military expedition, is quite clear on the subject:

The operation will not only offer great benefits to the country, on account of the very rich lands watered by the numerous rivers and streams which come down from the mountains ... but also on account of the security that would come to the present borders with the interception and cutting of the illicit traffic in cattle which since time immemorial the Indians have carried out with the provinces of southern Chile: Talca, Maule, Linares, Nuble,

Concepción, Arauco and Valdivia. . . .

I hold to the conviction that once this commerce, which makes Chilean finances rise or fall in direct proportion to the importance of the rustlers in Buenos Aires or other Argentinian provinces, has been cut off it would take away from the Indians the prime incentive that made them live as a constant threat to our wealth. . . . There are *caciques* who, *as foremen* of Chilean landowners, are entrusted with thousands of head of cattle which they return religiously after the winter. . . . At other times *they lease their lands* and the Chilean cattle raisers commonly live for long periods among them, without detriment to their interests. It is estimated that in that region alone – the northern part of Neuquén – some 20 to 30 thousand head of cattle winter yearly in the natural pastures of the Andes.

(Olascoaga 1976: 76–80).

The first contacts of the punitive expeditions with the Neuquén territory – minutely detailed in military dispatches – only confirmed this notable integration of the peoples on both side of the mountains. What is more, the dispatches show also that the agricultural practices of the Chilean Mapuches took hold in Neuquén, which was organized in relatively stable indigenous settlements, through the process of Araucanization already referred to.

Well into the nineteenth century, the incorporation of Argentina into the world market was shaping it as a country producing raw materials. Its agrarian economy which was essentially based on extensive cattle raising – through the *estancia* established as the productive capitalist unit and the casual *hacienda* which was the basis of the social organization of the Indian nation – began to disintegrate and eventually disappear. This encouraged increased rustling among the Indians as a means of economic survival, which in turn directly affected the interest of the Buenos Aires cattle raisers as the dominant sector of the new model of development.

The extension and consolidation of the internal frontier of the country became the key preoccupation of governments and the object of policies throughout the first half of the nineteenth century. A series of internal conflicts, including the war with Paraguay, had delayed and even paralysed the advance of the frontier with the Indians. But the perception of the Neuquén border economy as a permanent menace to the most productive sector of the country, linked to world markets, would lead to more drastic measures by that same sector. Following the so-called Desert Campaign, the deliberate subjugation of this elusive indigenous border economy was planned with the Neuquén territory as the theatre of operations. The region's rugged topography offered favourable conditions for the resistance of the last indigenous redoubts, which ultimately proved useless in the face of the technological superiority of the national army.

Upon the arrival of the military forces there was not a single white Argentinian settlement in Neuquén. As far as we know, only Chilean settlers shared the space with the Indians, as is attested by the settlement of Malbarco, mentioned and described in military dispatches. It is worth pointing out the level of complexity of this settlement, located in the extreme north-west of the territory, which was described at the time as a settlement of some 600 inhabitants of Indians, herders and Chilean settlers who leased lands from the *caciques* of the region (Olascoaga 1974: 78 and 368–9). Indians and Chileans maintained in Malbarco a particular way of life; the officials of the neighbouring country extended their *de facto* authority through the presence of civilian sub-delegates while recognizing the power base of the local *caciques* to lease their lands or to sign treaties tending to secure favourable treatment with persons or *haciendas* of the Chilean merchants or residents on the other side of the mountain range.

According to the same military sources there were settlements of sedentary or quasi-sedentary Indians in other parts of the territory with herds, enclosures and a semblance of agriculture which included irrigation by means of ditches. Furthermore, the known practice of transhumance – the seasonal movement of herds from the lowland winter camps to the high camps of summer – as the spatial characteristic of the animal husbandry of Neuquén, forces us to reject the common attribution of nomadism as the main characteristic of these peoples. However, the indigenous population's social and spatial relations using the verticality of the Andes Range as its principal axis was not overcome by the mere military occupation of this border region.

In fact, the mixture of sedentary and pastoral activity was common during the second half of the nineteenth century to the entire western region of Neuquén, particularly in the mountain and Piedmont areas, the topography of which favoured such activities. The eastern portion of the territory, with physical characteristics quite similar to those of the Patagonian Plateau, merely served as a transit point for the active commerce in cattle with Chile. It must not be forgotten that the geographic location of the territory as well as its topography also determined one of the key elements of the economic organization of these peoples: the movement of herds through the Andes Range. The indigenous populations of Neuquén constituted a fundamental element in the broad mercantile circuit that joined the cattle production of the pampa with the consumer markets of the Pacific.

In sum, these modes of indigenous social formation characterized the first stage in the occupation of regional space, with farmland and cattle-raising activity located essentially on the mountain slopes. Agriculture was tied to the internal consumption of the indigenous community and cattle raising was a basic element of an active commercial exchange with the Chilean cities and ports. The mountain and Piedmont areas of the Neuquén territory were already functioning as an integrated region centred around local towns (Chillán, Angol, Antuco) at this stage and as a hinterland for the principal Pacific ports at that latitude (Valdivia and Concepcion).

With the occupation of the territory by the military forces of Argentina in the 1880s, a fact no doubt aided by the simultaneous involvement of Chile in the Pacific War, the attempt was made to impose a territorial organization to support the new scheme of domination. In the interests of military security, it was decided to create towns and establish a capital – Chos Malal – as the political centre with maximum authority over this loosely integrated border region. However, structural characteristics of the dominant cattle business favoured the survival of old forms of social organization inherited from periods prior to the miliary conquest.

On the other hand, the incorporation of an indigenous region such as Neuquén into the national territory of Argentina brought with it the private appropriation of land as a productive resource. Dispossessing these communities of their natural conditions of production and transferring them to new owners provided the basis for a different social formation integrated into the national and international systems of the time. From that moment on, the different modes of capitalist accumulation of excess profits produced differentiated settlements which constitute the main topic of interest of this chapter.

THE RISE OF THE REGIONAL CATTLE-RAISING CYCLE (1879–1930)

The economic expansion of the country, based so far predominantly on cattle, demanded the incorporation of new lands to alleviate the pressure on the pasture lands in the Buenos Aires plains, while increasing the volume of production as a response to the European demand for wool and meat. The expansion of the internal settlement frontier also served the explicit objective of turning marginal lands over to the surplus cattle from more intensively occupied areas. After the military campaign of 1879, Patagonia was progressively incorporated into the national and international economic systems. The boom in sheep herding had initially affected only the territories on the Atlantic seaboard, and did not extend to Neuquén where pre-existing forms of socioeconomic organization persisted, in symbiosis with the core of the Chilean economy.

In this regard, the characteristics of the dominant economic structure of Chile must be borne in mind. In particular, the increased importance of agriculture during the latter half of the nineteenth century was partly due to the discovery of gold in California and Australia, where food supplies were still scarce, and to heightened demand from European countries (especially England). During the 1860s, Chilean agricultural production increased fourfold to become, together with copper, one of the most important export categories. Valparaíso became the principal port of the southern Pacific coast of the Americas.

The expansion of agricultural activities was reflected in the boom in the production of cereals prevalent in the southern provinces of Chile during the same period. For this reason, it is easy to infer how the transAndean trade from Neuquén could provide large quantities of meat – especially beef – without affecting land use in the more intensively cultivated central Chile. Thus, this transAndean cattle trade provided the Chilean economy with meat for internal consumption and with export commodities to meet world demand for tallow, leather and processed beef from other South American countries with Pacific ports such as Peru and Ecuador. According to Vicuña Mackena (Assadourian 1982: 57), the seventeenth century had been for Chile the 'century of tallow' and the city of Concepción – bear in mind its proximity to the Neuquén territory – produced large quantities of that product, destined for the markets of the Pacific. Later, the specialization in cereals already mentioned would spur the southward expansion of the agricultural frontier of the Chilean central valley, thus increasing the need for imported meat and its derivatives for local consumption and for export. Great quantities of cattle on the hoof (heavy but of poor quality) were required as raw material for conversion activities (tanneries, tallow candleries, soap factories, salting works) which dominated the Chilean urban economy at the time.

To these conditioning economic factors must be added physical characteristics of singular importance which made the Neuquén territory a functional hinterland for the Chilean economy. Because of its situation to the east of the Andes Range, Neuquén contained excellent physical conditions for satisfying the growing demand for meat products. I refer particularly to its ecological advantages: scattered forests, natural pastures adequate for cattle raising and transverse valleys which facilitated the movement from one side of the range to the other. Chile on the other hand possessed, at the same latitude, very dense forest areas little suited to cattle raising, with the sole exception of the valleys already taken up by agriculture.

While the main Argentinian cattle-ranching regions were improving breeds for meat destined for the European market, Neuquén was producing a more rustic type of cattle to meet the special demands of the Chilean markets. This situation was itself enhanced by the fact that Mendoza, the traditional supplier of cattle to Chile during that same period, was increasing its wine production and converting its pastures of alfalfa to vineyards.

Furthermore, the north-west region possessed an inherited infra-structure of roads which favoured the characteristic orientation of this border society. This region had formerly been the centre of Indian trails that went into Chile. An important number of the mountain passes made for rapid access to that transAndean region, aided by the greater permeability of the mountains in that sector of Neuquén, less rugged and cleared of forests at the crossing points.

These pre-existing transboundary networks in the west-central/north-west area of Neuquén explain why it was chosen for the establishment of the first stable capital in the territory, Chos Malal, in a geographical spot already preferred by the Indians as the hub of circulation and transit. Clear criteria of security and military defence determined the establishment of this regional capital at a time when an armed confrontation with Chile was feared. One must bear in mind the tense relations existing between Argentina and Chile towards the end of the last century.

At the turn of the present century, the broadening of the social organization of the territory began, mostly linked to the cattle activity, spreading to the whole of the mountain and Piedmont region. These private establishments occupied the lands in the southern territory opened up by the State following the campaigns to Nahuel Huapi and the Andes undertaken between 1881 and 1885. As a result of the consolidation of this cattle frontier, it was also decided to create new towns in the area, such as Junín de los Andes (1883) and San Martín de los Andes (1898), which later became service centres for the population of the surrounding areas. The better quality lands of the Neuquén Piedmont considered apt for cattle were opened up by ranchers, displacing the small goat raisers who had settled the area earlier on. The gradual and continued development of these lands – in accordance with the scale of exploitation of private lands – consequently gave rise to an important increase of cattle and sheep herds in the southern portion of the territory, as is evidenced by the successive census registers, particularly in the Departments of Aluminé, Huiliches, Lácar, Collón Curá and Catán Lil.

By 1930, the use of the land for cattle raising had been defined. By then, however, the world economic depression and the ensuing crisis in the activity produced notable changes. On the one hand, the first stage of effective occupation of the lands by private property had taken place in the south of the territory, thus determining the character of the area in direct relation to the economic importance of the new social actors linked to the regional cattle industry. On the other hand, the area occupied by people of limited resources dedicated to the raising of sheep and goats spread to the east of the territory, no doubt due to the dual effect of the continued increase in spontaneous migration of Chilean populations and the gradual displacing of the small breeders on to public or marginal lands. It should be made clear that following the military conquest of the territory, only the better lands were privatized on account of their productive capabilities but their effective occupation was slow. A demanding but limited regional cattle market served to restrict rapid and safe profits and therefore initially made Neuquén an area of scant attraction for outside capital. Its general lack of growth and wealth, the obvious neglect on the part of the national government and its weak demographic constitution effectively reflected that reality.

As a whole, little land was privatized in Neuquén (even today public lands take up more than 50 per cent of the province). This facilitated the spontaneous settlement of the population on public lands, generally Chileans and some direct descendants of the old owners of the land, now landless and considered 'intruders' into the new mode of production. Thus two distinct social groups related to the regional cattle industry were defined: the large producer-proprietors (with both cattle and sheep) located essentially in the southern zone, and the small breeders (mostly engaged in raising goats) occupying scattered public lands throughout the provincial interior. The middle-level producers were less representative. At the same time, the need to satisfy the basic needs of the rural population gave rise to the rapid formation of another relevant social sector: the merchants.

These were known as the *bolicheros*, owners of general stores scattered throughout the rural area, whose first representatives were the *mercachifles*, pedlars and travelling men, generally of Syrian and Lebanese origin. These merchants bartered general goods of Chilean origin for hides, wool and hair among the small- and middle-level producers, which they then sold in Chile.

Technological innovation during the period of this study was minimal, which confirms the predominant role played by traditional extensive cattle ranching. For instance, the use of barbed wire and the growing of forage was rare. Furthermore, the persistence in time of the traditional forms of commerce with Chile – practically monopolistic up to 1930 – and the final destination of the greatest part of the production (salting works, tanneries, tallow factories) did not demand great quality, thus discouraging the improvement of breeds and the use of barbed wire. The processing of raw materials was almost nonexistent in Neuquén, the sole exceptions being a salting plant for internal consumption and several rudimentary cheese factories of the cottage industry type. The spatial orientation of the economy was towards Chilean centres, where processing activities complemented the Neuquén cattle-breeding areas.

In effect, Neuquén made up the deficit in meat and other cattle-derived raw materials, as well as wool and goat hair, while the southern Chilean provinces almost exclusively specialized in agriculture. Chile, for its part, provided items of basic consumption: wine, sugar, beer, spirits, preserves, noodles, rice, paraffin, candles, soap, wood, linen goods, general merchandise and paper, tea, coffee, good-quality flour, etc.[4] Significantly, Neuquén often consumed manufactured goods whose raw material originated from the region, as in the case of soap and candles. The difference in freight charges was noteworthy in relation to those goods that came from Buenos Aires, Bahía Blanca or northern Patagonia.

Commerce was exclusively with Chilean markets (Valdivia, Temuco, Victoria, Los Angeles, Chillán, Concepción, etc.) so that the only currency in circulation was the Chilean Peso. This was the common situation until 1930, when the reorientation of the economy towards the rest of the Argentinian economy began. This was chiefly due to changes brought about at that time in market mechanisms and integration efforts by the central government in Buenos Aires. Any changes in frontier matters necessarily had serious consequences for the region. With increased political friction between Argentina and Chile, relations on the border deteriorated, and, as happened during the Pacific

161

War period in the 1880s, the use of Chilean currency in commerce tended to disappear. In response, there was a growing tendency among merchants to resort to barter or credit arrangements with their customers. The same occurred with customs restrictions which, always coinciding with periods of border conflicts with Chile, Argentina would set up and then later withdraw. When customs controls increased the effects were immediately felt at the regional level and across the entire border transect. Similarly, any change in the Chilean economy was immediately transmitted to the region.

The arrival of the southern railway at Confluencia (1902) and its later extension to Zapala (central zone) in 1913, is generally considered to be the first disruptive element in the exclusive commercial exchange with Chile in the whole of the mountain area. Nevertheless, it does not seem to have had a definite effect on the interior of the territory. The immediate relocation of the capital to the eastern slope – Neuquén, 1904 – caused Chos Malal to lose its political primacy, with a consequent decrease in population density. This progressively led to its transformation into a mere service centre for the rural area around it. The arrival of the railway in the border region and the erosion of the region's political position did not result in the automatic linkage of the mountain and Piedmont areas to Argentine markets. These areas maintained privileged commercial relations with Chile, staying free of the process of Argentine economic integration which affected in particular the eastern region of Confluencia.

The activities and modes of accumulation which have been described thus determined the pattern of settlement for the 1879–1930 period with its contingent socioeconomic consequences. The persistence of the tie with Chile as a feature inherited from the indigenous social formation determined, as we have seen, a major social organization in the western area of the territory which operated as a hinterland for Chilean urban centres. This fact remained unaltered in spite of the military occupation of the region and the imposition by the Argentine State of fixed boundaries and administrative centres of control. For these reasons, the only localities which manifested their full allegiance to the Argentinian State were answering political demands to reorganize the territory in some different form. However, these were often limited to a purely administrative character which did not alter existing social and economic relations. On the contrary, the social relations described in the first stages of historical evolution of Neuquén favoured, on the one hand, the rise of a few enterprises by large land owners and, on the other hand, the proliferation of small spontaneous settlements on public lands, in areas appropriate for the winter grazing of cattle. The latter were destined to suffer, as we shall see, marked changes in their role with the onset of the crisis that followed the closing of the boundary with Chile.

THE BORDER CRISIS AND ITS CONSEQUENCES

Beginning in the 1930s, signs of the great break-up of the mountain region were apparent as the suppression of free commerce through Chilean and Argentinian state policies seriously affected the spatial articulation of the economy.

Due to the effects of the international crisis and the subsequent adoption of protectionist measures, more severe customs controls were put in place by 1930. In addition to the imposition by the Chilean Government of a high tax on the import of cattle from the other side of the Andes, a further 10 per cent duty on imported merchandise was enforced by the Argentine Government in 1931.[5] Matters became worse with the application of the protectionist agreements of the same year on exchange controls by which merchants and cattle producers were required to stop at the border to meet tax requirements before they could do business.[6] This clearly altered the *modus operandi* of this transAndean economy, where transactions between Chilean and Argentine merchants had traditionally been of a spontaneous and informal nature.

These state fiscal policies paralysed commercial transactions, equally affecting the whole of the mountain and Piedmont regions of Neuquén which had traditionally functioned as a region responding to the demands of Chilean centres and ports. Years later, in the 1940s, the situation took a turn for the worse when the industrialization phase of the Argentine national economy imposed greater customs barriers, thus marking the end of an era in the economic history of the Neuquén interior.[7]

Both countries, at different times and for different reasons inherent in their own historical circumstances and the international situation, consolidated their territorial organization through the fixing of a commercial frontier with varying social consequences. The possibility of modifying behaviour and the adoption of different economic alternatives by the productive sectors, whether large-scale producers or small- and medium-sized units, depended on the varying positions of power of each in relation to the mechanisms of accumulation in the economy. Thus, the possibilities for accumulation of the small- and medium-sized producers gradually dwindled as the exchange with Chile, which had been carried out directly up until 1930, was interrupted.

The crisis brought about by the disruption of the commercial network is reflected in the total numbers of cattle, which decreased constantly from the 1940s onwards, only to recover in some areas in the mid-1970s when the recuperation of the regional cattle industry began.[8]

With regard to the changes mentioned, it could be argued that the reorientation of the commercial networks in the 1930s and 1940s, and the degradation of soils and vegetation as a consequence of over-population, were the direct causes of the productive changes in the north-west region. In the south-west and south-east the loss of yield in wool-producing animals was the determining factor in the change of activity. And at the same time, the increase in the goat population, or the reverse, the replacement of wool-bearing animals by bovine herds, was directly linked to the alternatives adopted, in the face of crisis, for accumulation among cattle ranchers. This determined the difference between the southern zone with its beef-cattle specialization and the northern zone with its gradual increase in goat populations, as well as the effects in the patterns of settlement and depopulation of the rural areas of the province.

As long as the strength of the activity in the region was breeding, cattle raising maintained extensive production systems without substantially modifying the technological forms of cattle management common during the prior stage. There is evidence of an increased tendency towards improving the strains of cattle and sheep in the large productive enterprises as a way of ensuring the possibilities of placing these in the national market once the commercial link with Chile was interrupted.

The paralysis of commerce with Chile affected equally all the productive strata but, obviously, resulted in more serious socioeconomic consequences for the small producer, the social actor of greatest relevance in the north-west, central and eastern regions of the territory. Although less seriously, given the scale of production of the more representative businesses of the area, the damages caused by the crisis also affected the larger producers characteristic of the southern zone. These firms, essentially geared towards beef production, were seriously hurt by the tax on imports exacted by Chilean customs and the low prices on the national markets.

In the face of the critical situation described above, the only possible alternative for local producers seems to have been the entry into the national market, a complicated matter if we bear in mind, on the one hand, the devaluation of the herds as a consequence of the international crisis and, on the other hand, the lack of an integrated transport system with the major Argentine markets (Bandieri 1983: 45). Thus the only viable way of channelling cattle from the interior of Neuquén toward the markets of the Atlantic was shipping the herds by railway, especially from Zapala in the centre of the province. But such an operation incurred prohibitively high freight costs which substantially lowered profits for the producer.[9] Sources indicate an important increase, starting in 1930, in the export of herds and other products toward the markets of Bahía Blanca and Buenos Aires.[10] This supports the idea that the imposition of customs controls with Chile redirected a large proportion of the trade from the interior of the territory toward centres of national commerce, thus transforming the traditional economic relationships in the region.

Both the interruption of commerce with Chile and the high railway freight rates affected all cattle producers similarly. However, the large producers had access to political solutions which the system itself offered them. In 1933, in response to repeated requests, the Nación Bank agreed to suspend for one year payments on debts, letting current interests accumulate 'on account of the economic breakdown which the Piedmont region is suffering because of the paralysis of commerce with Chile and the depreciation of the cattle and fruits of the country'.[11] Official efforts were also made with the administration of the southern railway to obtain lower freight rates for the transport of cattle from Zapala to the wintering areas of Buenos Aires, a request to which the company acceded by establishing special rates.[12] It is clear that these measures favoured only the large producers who had access to the credit system and those who sold cattle for wintering in Buenos Aires or who also owned property in the Province of Buenos Aires where they could finish the fattening process, as was becoming common among many of the cattlemen of the south.

By 1940, very little foreign money was entering Chile, and a few years later exports were at a standstill. Indeed, as of 1945 the Argentine Government imposed strict controls on international traffic with Chile. Rigid rules of the Central Bank regulated export and import and required a prior deposit in foreign exchange related to the value of the products to be exported, all of which brought down the traditional cattle trade of Neuquén. Dependency on the national market for supplies deepened, and by the end of the 1940s most consumer goods and capital came from Buenos Aires or Bahía Blanca, with a consequent increase in costs.

The severe control over the traditional commercial traffic with Chile imposed impossible conditions on the small- and medium-sized producers. It is clear that beginning in the 1930s, and certainly by the 1940s, the complete integration of the region's cattle production into the national market was accomplished. This occurred in spite of an enduring age-old trade network with Chile which was maintained as long as there was the real possibility of trading with that country. Of course, it is worth mentioning that smuggling continued as a viable, albeit risky, alternative given the presence of greater border control elements. Obviously, this practice did not constitute a solution to the problem, but it often served as a timely alternative to the extent that it allowed the illegal trading of small amounts of cattle with Chile, although in much smaller numbers than had previously been possible. The political and customs controls definitely put an end to the regular placing of large herds of cattle on to Chilean markets.

When the traditional markets were severed different producing firms adapted their strategies according to the market power of the firm at the moment of change. Thus, the large enterprises – predominantly engaged in cattle production, and to a lesser degree in sheep raising – who were able to dispense with commercial intermediaries, located primarily in the central and southern mountain areas, found alternatives for obtaining a rate of return that would at least allow them to continue in the activity in spite of the crisis of the traditional system of commercialization. On the other hand, smaller productive units often depended on a chain of intermediaries in the process of commercialization, thus reducing further their opportunities for accumulation and frequently driving them out of business.

During this process the north-west region was without doubt the hardest hit. As was to be expected, as of 1930 the circulation of Chilean currency came to an end. The mercantile sectors sought adjustments in other avenues of circulation – Mendoza, Zapala, General Roca. Small producers faced with few or no possibilities of productive relocation became increasingly marginalized and lost their independent forms of commercialization already described. With the devaluation of the herds and their subproducts (wool, hides, hair, etc.) and the paralysis of commerce with Chile, the small producers were forced to offer their already devalued products to the local market in bartered exchange for Argentine consumer goods which were markedly dearer. Precapitalist social relations of production which had persisted up to this point were thus intensified, which is explained by the internal logic of the system as businessmen upon entering a wider circuit where the chain of intermediaries is greater would need the offer of producers with restricted power of negotiation to allow them to maintain a margin of profit. In the 1930s and 1940s these mercantile sectors became the only possible access to the national market for small producers. Thus they broadened their base in search of greater profits and, together with the large entrepreneurs, made up the local power structures. As part of the local bourgeoisie, they increased their political power with the establishment of provinces in the territory around the mid-1950s.

CONCLUSIONS

The impact of national integration policies on local border economies is a subject of current interest. In this study, a particular transboundary economy between Argentina and Chile has been analysed historically. The border economy linking Neuquén to Chilean markets endured for centuries until the implementation of customs and border controls between the neighbouring states dramatically changed the social and economic systems in this mountainous region. The interruption of free trade with Chile by the fiscal policies enforced by both national administrations during the 1930s and 1940s destroyed the principal commercial relations that had defined an enduring social organization and its spatial expression. These measures triggered the crisis in the regional cattle activity which would become worse in later years.

Faced with the crisis, social actors related to the border trade had different alternatives directly related to their possibilities of accumulation and their relative weight in power relations at the regional level.

As a consequence of the reorientation of the local economy, the large producers and entrepreneurs of Neuquén saw their possibilities for accumulation enhanced and went on to occupy an important place in the power structures at the regional level, becoming one of the outstanding elements of the local bourgeoisie. On the other hand, small productive units which had thrived from the spontaneous occupation of public lands in the central and north-western areas of Neuquén and from direct trade with Chile, enjoying while it lasted a relative economic independence, suddenly saw their modes of commerce seriously affected. The possibility of selling their cattle locally depended on successive commercial intermediaries who became the indispensable link to the market for the small productive units. Thus these small producers had but two alternatives: to stay in place to assure subsistence, or migrate to the urban centres to become part of the labour force in other types of activity. This explains the ongoing depopulation of rural Neuquén which began in the 1940s. The survival of the subsistence economies and the impossibility of generating surplus profits tended to increase the degree of marginality in Neuquén, and has prevented economic change ever since.

NOTES

1 By 1879, the year which saw the so-called Desert Campaign or the definitive conquest by the whites of the Patagonia territory, the process of Araucanization was complete in the entire area.

2 With the spread of cattle and horses introduced by the Spanish which, in the freedom of the Argentinian plains, became common property, the economic activity and the lifestyle of the people was profoundly altered. The decrease in the numbers of 'wild' animals through the indiscriminate slaughter of the herds and the subsequent establishment – well into the first half of the nineteenth century – of the cattle ranches as the productive unit made it more difficult to acquire those goods that had become the basis of a way of life: cattle, for consumption and for trade. This brought about the increased recourse to rustling as a way of advancing over the plains of the pampa and of obtaining that formerly common property which had now become private.

3 The conclusive military expedition against the Indians, called 'The Desert Campaign', reflects in its very name the ideological conception of the dominant sectors that carried it out. In fact, the term 'desert' is here used in a social rather than an ecological sense and is synonymous with barbarity or, what amounts to the same thing, the absence of civilization.

4 Report of the Chilean Consul General in the Republic of Argentina, in *Boletin del Ministerio de Relaciones Exteriores*, Santiago de Chile, First Semester, 1902.

5 *Anales de Legislacíon Argentina*, 1920–1940: 253–4.

6 Provincial Historical Archive (1933), Note of January 11, 1933, Libro Copiador de Notas al Ministerio de Interior, folio 82, October 10, 1932 to September, 1935.

7 As of 1945, the Argentine Government imposed severe controls on international traffic with Chile. Rigid rules of the Central Bank regulated imports and exports, requiring prior deposit of foreign exchange related to the value of goods to be exported. The effects of these measures on the deteriorating traditional cattle industry of Neuquén can be seen in *Libro Histórico No. 1* of the Escuela Nacional No. 15 of Chos Malal, Neuquén, founded in 1887, now Escuela de la Frontera No. 3; MS, 1946, folio 32.

8 These signs of recuperation could be attributed to, among other things: the initiation of the 'Shearing Program and Commercialization of Wool of the Province of Neuquén' (1975); the establishment of a sanitary cordon to keep the region to the south of the Colorado River free of hoof-and-mouth disease (1976); and the different cooperative efforts of commercialization undertaken by the small- and medium-sized cattle producers of the interior of the province (Bandieri 1988: 226–30).

9 Provincial Historical Archive, Libro Copiador de Notas al Ministerio de Interior, report dated February 1933, folio 175. Similar concepts are repeated in the Memoir presented to the national government by Governor E. Pilotto, 1934, folio 105.

10 Provincial Historical Archive, Libro Copiador de Notas al Ministerio de Interior, report dated February 1933, folio 175. Similar concepts are repeated in the Memorandum presented to the national government by Governor E. Pilotto, 1934, folio 105.

11 Provincial Historical Archive, Libro Copiador Varios, March 4, 1933 to October 1, 1933, Note of June 10, 1933.

12 Ibid., Note of September 19, 1933.

BIBLIOGRAPHY

Assadourian, C.S. (1982) *El sistema de la economía regional. mercado interno, regiones y espacio económico*, Lima: IEP.

Bandieri, S. (1983) 'Evolución histórica de la ganadería y su distribución por zona', in CALF-Gráfica Modelo, *Neuquén, un siglo de historia*, Neuquén.

―― (1988) 'Condicionantes históricos del asentamiento humano en Neuquén: consecuencias socioeconómicas', Unpublished manuscript presented as a report to the CONICET, Buenos Aires.

Bandieri, S. and Angelini, M.C. (1982) 'Evolución de las existencias ganaderas neuquinas entre 1895 y 1937' in *Actas de las IV Jornadas de Historia Económica de Argentina*, Proceedings of Argentinian Historians, Córdoba: Univeristy of Córdoba.

Edelman, G.C. (1925) *Informe sobre ganadería, agricultura, comercio, industria, etc. Territorio del Neuquén y parte de Río Negro*, Buenos Aires: Taller Gráfico Neumann y Cía.

Giusti, V. (1984) 'Sistema de comercialización de ganado en pie en la Provincia de Neuquén', Mimeo., Proyecto de Cooperación Técnica para el desarrollo de la región Sur: O.E.A.

de Monsalvo, D. (1980) 'Actividad caprina en el Norte neuquino', Unpublished thesis, Departmento de Geografía, Facultad de Humanidades, Universidad de Neuquén-Comahue.

Olascoaga, M.J. (1974) *Estudio Topográfico de La Pampa y Río Negro*, Buenos Aires: Euleba.

Siri, A.F. (1983) 'Sistema Provincial de comercialización de productos ganaderos', Mimeo., Technical Report: O.A.S.

UTS-BAHN (OAS Consultants) (1983) 'Desarollo predial de la producción ganadera en el Area Norte-Provincia del Neuquén', Mimeo., Neuquén: February.

Yazman, J.A. (1983) *Perfil de un plan de mejoramiento de la producción caprina en la zona Norte de la Provincia de Neuquén*, Mimeo., Neuquén.

11

THE BOLIVIAN MARITIME ASPIRATION TO CHILEAN POLITICAL SPACE

Towards a non-territorial solution

Monica Gangas Geisse and Hernan Santis Arenas

INTRODUCTION

The subject presented here is the preliminary product of comprehensive research on Bolivian maritime aspirations to Chilean political space. These territorial aspirations have constituted the purpose and political objective of generations of Bolivian politicians and intellectuals ·in pursuit of access to the coast and influence in the Pacific Ocean.

In order to characterize the Bolivian political objective and purpose, this chapter is divided into three sub-topics or sections. First, the Chilean–Bolivian territorial relations are analysed in order to introduce the subject. Second, several hypotheses which explain Bolivian pressure upon Chilean political space are reviewed. Finally, some options for non-territorial solutions which may diminish or eliminate the pressure exerted by the Bolivian aspirations to the Chilean political space are reviewed.

A brief review of the territorial–political relations between both states is necessary in order to understand the Bolivian maritime aspiration to Chilean political space. Usually, it is thought that the origin of the issue is the economic and territorial expansion of Chile over the Atacama Desert, located between the Loa River and Latitude 26° South, and between the Pacific Ocean coast and the high peaks of the Andes.

However, there is a more profound and real reason for the Bolivian maritime aspiration which had previous historical manifestations as a political–territorial objective of the emerging Bolivian Republic during the first decades of the nineteenth century.

POLITICAL AND TERRITORIAL RELATIONS
BETWEEN CHILE AND BOLIVIA

Until 1810, the territorial and political relations of the Chilean and Bolivian political systems were under the control of the Spanish colonial system in South America. Although both provinces were ruled by the King of Spain and were subjected to the Hispanic political and administrative system, they belonged to two distinct jurisdictions. The land and inhabitants of the Chilean realm constituted the *Capitanía General y Gobernación del Reino de Chile*, whereas the land and inhabitants of Upper Peru province constituted the *Audiencia de Los Charcas* (see Figure 11.1).

The *Audiencia* of Charcas was administratively subjected to the Peruvian Viceroyalty, while Chile was directly linked to the Iberian metropolis. The political liberation movements of the nineteenth century, which led to the emancipation of independent states, began in Upper Peru (1809). However, Chilean society was the first to obtain its political independence (1818); meanwhile, the Bolivar Republic – later known as Bolivia – was only constituted as a republic in 1825.

During the colonial period (lasting from the sixteenth to the nineteenth century), the Upper Peruvian maritime aspiration emerged due to the importance of the mining of precious metals in that province, and the need for an accessible outlet to the Pacific Ocean. Since the seventeenth century, the export of mining products had been through the seaports of Arica (Lower Peru) and Buenos Aires (Argentina). These economic flows partially accounted for the tendency of entrepreneurs, in 1825, to form associations with Lower Peru, and with the *Provincias Unidas del Río de la Plata* (Buenos Aires), or to proclaim their political autonomy, in the context of the wars of liberation.

In 1825, the Chilean Plenipotentiary Minister in England, Mariano Egaña, in a letter sent to his government officers in Santiago, described the Upper Peruvian options from the Chilean perspective. Egaña thought that the independence and political autonomy of Upper Peru meant the establishment of 'a balanced centre' which would guarantee Chilean political independence, while implying the non-existence of alliances between its territorial neighbours.

The idea of Upper Peru becoming a political centre to provide a political equilibrium among the countries in the region (Peru, Bolivia, Argentina and Chile) materialized by 1836 through the political action of Minister Diego Portales. By that year, Andrés de Santa Cruz, President of Bolivia, succeeded in securing the Peruvian–Bolivian

171

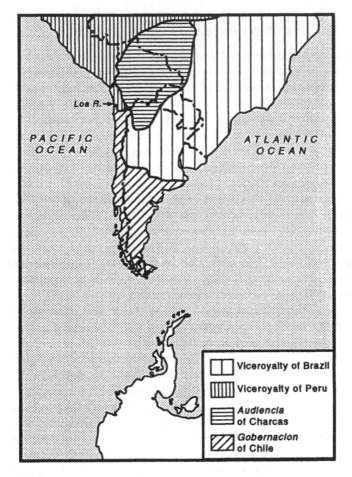

Figure 11.1 Political administrative territories of Hispanic kingdoms of Peru, Chile and *Audiencia de Charcas*

Confederation Agreement, a treaty which defined north Peruvian, south Peruvian and Bolivian relations. However, according to Portales such a treaty was a menace to Chilean political independence and to the security of the Chilean maritime trade in the south-east Pacific Ocean.

According to the political and diplomatic archival material on Peruvian–Bolivian relations, it is clear that Bolivia, in 1826, along with its first attempt to create the 'Confederación Bolivariana', obtained an agreement from the Lower Peruvian Plenipotentiary for a territorial exchange treaty. Through the Chuquisaca Treaty (1826), Bolivia would have received the Lower Peruvian territories between the Sama River and the Loa River. In return Peru would have received Bolivian territories in the province of Apolobamba or Caupolicán, including the city of Copacabana high in the Andes. The Peruvian Government deprived its Ambassador and Minister of Foreign Affairs of authority, and did not ratify the territorial exchange treaty (see Figure 11.2).

However, at the request of Bolivian political leaders, both General Bolívar and General J.A. de Sucre agreed in 1825 to transfer the seaport of Cobija, at the southern confluence of the Loa River and the sea, to

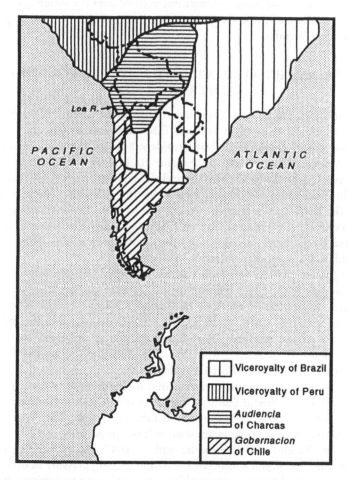

Figure 11.2 Political administrative territories of Peru (Low Peru), *Audiencia de Charcas* (High Peru) and Chile under the 1826 Chuquisaca Treaty

the nascent Upper Peruvian Republic. In those days, Chilean authorities, who had jurisdiction over the Atacama coastal region, did not protest at the breaking up of Chilean territorial integrity, partly due to lack of knowledge of the terms of *uti possidetis jure* established in 1810.

The result of the conflict between Chile and the Peruvian–Bolivian Confederation (1836–9), which favoured the former, did not involve any sort of territorial arrangements. The Chilean Government was interested in the dissolution of the Peruvian–Bolivian alliance and the feasibility of having adequate maritime control over the shipping routes of the south-east Pacific Ocean, in accordance with the economic plan of Minister Portales.

BOUNDARY LITIGATIONS AND TERRITORIAL CONTROVERSIES

During 1842, the Bolivian Government appointed Casimiro Olañeta as Plenipotentiary Minister in Santiago. The main task of this diplomatic official was to persuade Chilean President Manuel Bulnes to convince the Peruvian Government to transfer to Bolivia a seaport located between Quebrada de Camarones and the Loa River, probably Pisagua on the Atacama coast (Abecia Baldevieso 1979, I: 498). The idea was that the Bolivian Government would bear in mind the obligations that the Peruvian Government had to Chile regarding the expenses contracted by the latter, deriving from the Confederation War. Thus, Peru would cede a seaport to Bolivia in exchange.

In January 1843, after failing in his attempts, Olañeta submitted a protest note to the Chilean Government. By means of this note he asked for the abrogation of the Law of Natural Guano Deposits (October 1842), many of which, according to his statements, were located in Bolivian coastal territory. For this purpose, he indicated that the Despoblado de Atacama, from the Loa River to the Salado River, belonged to Bolivian territorial jurisdiction.

The Chilean Government replied that it could not alter the existing laws without a careful study of the titles that Chile or Bolivia could show for territorial rights in the Atacama Desert. The boundary controversy between both states was, at least temporarily, settled.

From the end of 1825 to the end of 1842, the Upper Peruvian authorities and political officials exerted increasing pressure in order to obtain a coastal territory, through direct or indirect negotiations, from Peru. Beginning in January 1843, all the pressure for the attainment of a coastal territory which would secure Bolivian maritime aspirations was reoriented towards Chile.

THE LONG ROAD TO A SOLUTION

The settlement of the controversy regarding the boundaries and territories of the Despoblado de Atacama lasted over a sixty-one-year period (1843–1904), including stages of political transactions, warlike conflicts, temporary cessations of hostilities and a final peace treaty.

By 1845, the Chilean Government was able to gather all the geographical and cartographical data which supported its statements concerning the Chilean–Peruvian boundaries in the Loa River (Latitude 21° 48′ South). In 1847 the Bolivian Government rejected these statements, resorting to a linguistic argument concerning the origins of toponyms in the region. In 1858, the Bolivian Government insisted again on the Salado River boundary, citing the text of the political constitution of the Chilean Republic in which it was stated that the national territory extended from the Atacama Desert.

In 1863, the Bolivian Government issued a declaration of war against the Chilean Government. In Santiago, news of the war was received belatedly. In the midst of this potentially dangerous situation, Chilean–Bolivian diplomatic relations were renewed, facilitating the negotiation of a boundary treaty (1866). As a consequence of the war with Spain, the Chilean Government had to rebuild its naval power, to reconstruct the seaport of Valparaíso, to strengthen various coastal ports and to reorganize its commercial fleet. All this led to a political transaction concerning the boundary in the Atacama Desert, which was fixed at Latitude 24° South. In accordance with its fiscal policy, joint control was established over an area between Latitudes 23° and 25° South for the purposes of levying taxes on the export of mining products.

In 1872 Chilean–Bolivian negotiations were renewed, deriving from the Lindsay–Corral Protocol, whose objective was to lay down the bases for a final agreement which could resolve the ongoing ambiguities of the treaty of 1866 and pave the way for subsequent negotiation and a new treaty. Finally, in 1874, a new boundary treaty was signed. Joint fiscal control was abandoned and Latitude 24° South was established as the boundary. Bolivia agreed not to increase rates and taxes to Chilean entrepreneurs operating in the territory of Atacama during a twenty-five-year period (see Figure 11.3).

In 1878, the Bolivian Government issued a law revoking the exemption of duties previously agreed with Chilean companies. Chile

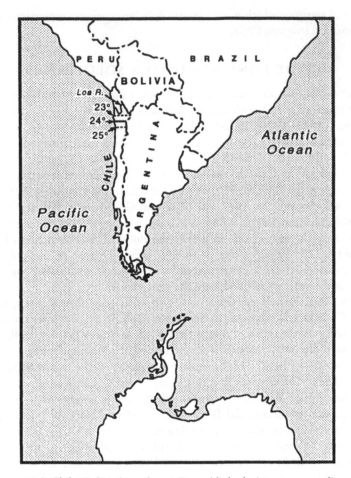

Figure 11.3 Chile–Bolivia boundary at Despoblado de Atacama according to
treaties of 1866 and 1874

complained about this and demanded Bolivia fulfil the international
treaty. Bolivia offered to resort to the arbitration procedures detailed in
the treaty. In February 1879, however, Bolivia placed an embargo on
goods produced by Chilean companies because they did not pay
rates and taxes. The Chilean armed forces invaded the territory of
Antofagasta, taking for granted that Bolivia had unilaterally violated the
Treaty of 1874. In March, Bolivia declared war on Chile. By April 1879,
it was known that Bolivia had an offensive–defensive alliance with Peru,
which forced Chile to declare war on both of them.

It is clear that the war on land and at sea was favourable to Chile. Bolivia capitulated in 1880, and in 1883 Peru agreed on a Treaty of Peace and Friendship with Chile. The following year Bolivia signed the Treaty for the Temporary Cessation of Hostilities (1884).

On October 20, 1904, upon acceptance of the Treaty of Peace and Friendship, negotiations for which were started in April 1902, peaceful relations were re-established. Bolivia recognized the absolute and perpetual domain of Chile over the territories between the Loa River and Latitude 24° South, encompassing the coastline and boundary stated in the 1884 Treaty of Temporary Cessation of Hostilities.

BOLIVIAN POLICY AFTER 1904

In 1910, the Bolivian Chancellor, Daniel Sánchez Bustamante, asked Chile and Peru for an outlet to the sea. This proposal expressed the position of the *practicistas*, that is those Bolivians who longed for a seaport which would be possible only through the surrendering of the Peruvian territories of Tacna and Arica, under Chilean jurisdiction since 1883. The Bolivian opposition to such a solution came from the *reivindicacionistas* who uncompromisingly claimed the Chilean seaport of Antofagasta.

In 1919, the Bolivian Ambassador in Paris and London, Ismael Montes, approached the French Government in order to present a legal case by which both Tacna and Arica would be transferred to Bolivia. The following year the Republican Party came to power with a popular policy for a 'maritime replevy'. In November of that same year, Bolivia presented a motion to the League of Nations demanding the examination of the 1904 Treaty. In 1921, the League of Nations declared that such a demand was inadmissible. In 1923, the Bolivian Government asked Chile for a re-examination of the 1904 Treaty. Even in 1926 and 1927, Bolivia still favoured one of the formulas proposed by the American arbitrator, Frank B. Kellogg, concerning the controversy over Tacna and Arica, between Chile and Peru. The idea was that such territory be ceded to Bolivia, with the latter paying an indemnity to Chile and Peru.

In 1943, Bolivia initiated a campaign before the US State Department for a review of the treaty. In 1945, it intended to include in the text of the United Nations Charter a proviso that could support its thesis for the analysis of the treaty. In 1950, initiatives in favour of granting Bolivia its own outlet to the sea arose in the context of a project over the exploitation of the waters of Lake Titicaca.

Notwithstanding, the negotiations failed. During the second administration of Victor Paz Estenssoro (1964–5), a campaign of vindication was renewed both nationally and internationally. Between 1962 and 1975, Bolivia severed its diplomatic relations with Chile; the stated reason was the controversy over the exploitation of the waters of the Lauca River.

In February 1975, the Pinochet administration proposed to have a meeting with the Bolivian president, Hugo Banzer, in order to exchange opinions concerning affairs of common interest for both countries. In August of the same year, after the resumption of diplomatic relations, the Bolivian Embassy in Santiago presented an *aide-mémoire* which included the request for the transfer of a sector of the northern coastline of Chilean territory. The negotiation failed because Peru, which Chile should have consulted according to the 1929 Treaty, proposed a completely different scheme. In 1978, diplomatic relations were once again broken between Bolivia and Chile (see Figure 11.4).

The failure of these negotiations led Bolivia once again to expose its land-locked condition at an international level. In 1979, it stated its case before the Economic Commission for Latin America and the Caribbean and in the General Assembly of the Organization of American States (OAS). In 1987, Bolivia tried to obtain some territorial transfers from the Pinochet Administration by capitalizing on the relationship between the Chilean Chancellor and its Consul General in Chile.

CHILEAN POLICY BETWEEN 1904–29

According to the Treaty of 1904, Chile was compelled to build a railway to connect Arica and La Paz, transferring the property of the Bolivian section to Bolivia. Likewise, guarantees for developing Bolivia's national railway system were made. The Bolivian Government received credits from several international sources. Bolivia was granted extensive and free rights for commercial traffic throughout Chilean territory and seaports on the Pacific Ocean. The establishment of Bolivian customs agencies in Chilean ports was permitted. And finally, rail freight fees for Chilean mining and manufactured products bound for Bolivia were also reduced.

These favourable terms were established by several bilateral agreements, including the 1906 protocol concerning the forfeit of customs duties, the telegraphic convention of 1906, the agreements on railroads guarantees of 1907 and 1908 and the commercial convention of 1912, among others.

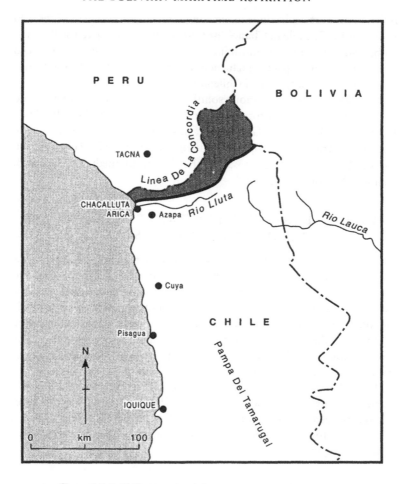

Figure 11.4 Chilean territorial suggestions to Bolivia, 1975

In January 1920, the Chilean ambassador in La Paz, Emilio Bello Codesido, formally stated that Chile was ready to look for the possibility of a Bolivian outlet to the sea of their own. This included transferring to Bolivia a significant portion of territories to the north of Arica, including therein the railway which had been subject to referendum stated in the 1883 Treaty. In 1921, Chilean President Arturo Alessandri stated, upon the arrival of a new Bolivian ambassador, that he could not accept a revision of the treaty. Notwithstanding, he was favourable to new negotiations concerning

Bolivian seaport aspirations in exchange for proper compensations. On February 1927, Chilean Chancellor Luis Izquierdo wrote to the Bolivian ambassador in Santiago, informing him that the government had agreed to consider a new pact which 'neither modified the 1904 treaty nor interrupted the continuity of Chilean territory'.

In 1926, among the various solutions discussed between Chile and Peru concerning the territory between the Sama River and Quebrada de Camarones, the Chilean Government hesitated between the division of Tacna and Arica and the option of giving arms to Bolivia. This situation led to the arbitration proposal of the American mediator, Frank B. Kellogg (November 30, 1926).

THE CHILEAN–PERUVIAN TREATY OF 1929 AND THE END OF TERRITORIAL SOLUTIONS

In 1929, Chile and Peru finally reached a settlement of the controversy over Tacna and Arica with a definite boundary dividing the disputed territory. However, the treaty had a protocol which stated that the governments of Chile and Peru could not transfer territories – partially or completely – to a third party without a previous agreement between them. According to the treaty, the area remained under their respective sovereignties and they would not be able to build new international railways through them without prior consultation.

In 1976, the Chilean Government suggested to Peru the creation of 'a territorial corridor' for Bolivia between Chile and Peru, but the latter offered no response to the idea. Instead, Peru proposed the creation of a tri-state territory and the internationalization of the seaport and town of Arica (see Figure 11.5).

BOLIVIAN POLITICAL PRESSURE UPON THE NORTHERN AREA

Bolivia's aspirations to a seaport, as the 1904 treaty began to be revised, led to international pressure on Chile. Bolivia has laboriously sought a review of the boundary agreement resorting to several direct arguments with the Chilean Government, attempting to profit from the territorial litigations by giving the matter an international dimension. The international political image of Chile deteriorated as a result of these facts. The government has allocated substantial state resources to defend its northern border, implying the permanent upkeep of military technology, thus consuming public resources for military expenses. The inter-

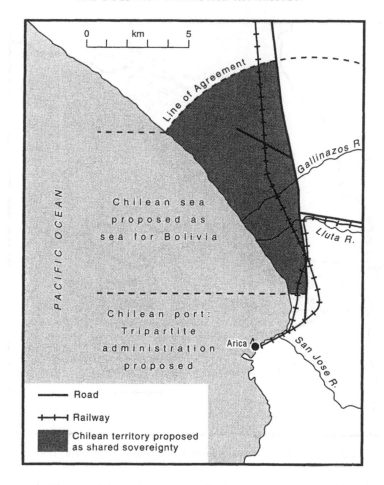

Figure 11.5 Peruvian suggestion to Chilean consultation about Bolivian corridor, 1976

national consequence of this policy has given Chile the image of a militaristic and belligerent country.

The juridical doctrine of the intangibility of treaties has compelled several governments, who sought to help in solving Bolivia's land-locked condition, to propose territorial exchange systems, but the Bolivian Government has systematically refused to collaborate with these initiatives.

Over the last eighty-seven years (1904–91), the different Chilean governmental administrations have exempted Bolivian commercial

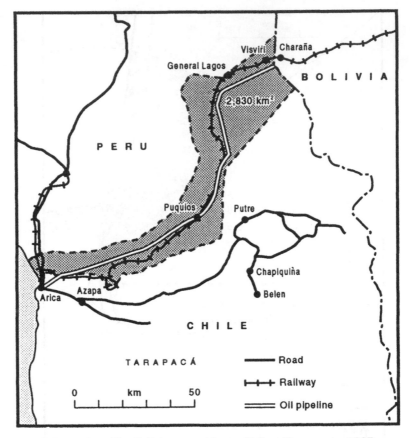

Figure 11.6 First Bolivian exposition to Chilean Government, 1987

traffic through Chilean territory and seaports from taxation (see Figure 11.6). Over the last decade, several complementary bilateral economic instruments have been proposed. However, Bolivian authorities have chosen to keep diplomatic relations with Chile at a very low level, insisting instead on making the problem international.

Therefore, Chilean public opinion has turned, over the last years, to opposition to any territorial solution concerning Bolivian maritime aspirations.

MAIN HYPOTHESIS

In the course of this research, several key questions arose in the debate over Bolivia's maritime prerogative. Historically, that fact has forced the Chilean political system to create a military organization in the face of Bolivian pretensions, and the more silent Peruvian pretensions over Arica and the territory of Tarapacá (see Figures 11.6 and 11.7).

The existence of two neighbouring political–territorial entities with a desire for replevying suggests on the one hand a hypothesis of conflict between the states and on the other a hypothesis of alliances between them. The military dimension of these hypotheses cannot be under-estimated, leading to the tense relations between the three states. The

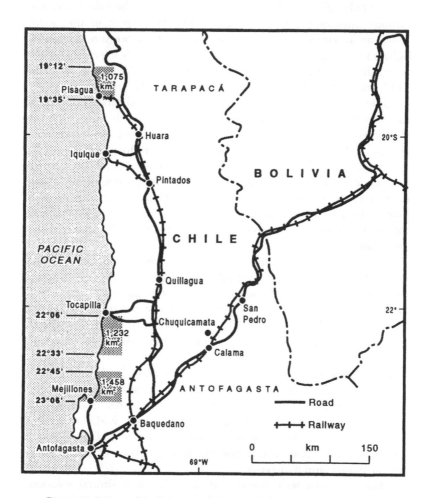

Figure 11.7 Second Bolivian exposition to Chilean Government, 1987

183

implicit or explicit defence policy of the Chilean state is understood, to a great extent, by the military development in relation to defensive aims, in an eventual northern theatre. Late in the twentieth century, the necessary military development to support security not only involves the quantitative increase of the staff of land, navy and air forces, but also makes it necessary to supply them with armaments and adequate logistic support for the defensive role they will be performing.

Following this approach, guided by the political–regional aim of keeping a 'balanced centre' – the thesis of M. Egaña (1825) – the Chilean political system has often been considered to be militaristic. The obvious question is, then, whether Chile takes part in the South American armaments race merely as a result of the pressure exerted by Bolivia in stating its maritime aspirations, historically expressed by the strong desire for having a territorial enclave on the Peruvian coast, until 1842, and on the Chilean coast from 1843 up to the present.

Chilean military development and the associated modernization of the defensive means of the political system in the northern area, affects the annual national budget. It is obvious that public expense on defence lessens or delays the possibilities to invest in productive areas of the economy, thus reducing the necessary attention to social services like housing, education and health. The cost for tax payers, in direct and indirect taxes, is high. On the one hand, there is the obligation to finance the defensive military force and, on the other, the state must pay for the free Bolivian commercial traffic throughout the Chilean territory and seaports.

The northern area of Chilean territory is rich and possesses a wide range of metallic and non-metallic mineral deposits. Along with that, the existence of soils is known which, even though arid, have proved historically and experimentally at present, to be potentially fertile for intensive agricultural exploitation. Nevertheless, the development of mining and agriculture are severely limited by the absence of water and power plants.

The experience of Chilean–Bolivian negotiations clearly shows that a territorial solution is impossible. Bolivia, on one side, insists on reviewing the treaty of 1904 and Chile, on the other, reiterates the principle of intangibility of international agreements. Without this firm Chilean policy, a complex set of judicial structures would crumble in a very short time, thus threatening the very existence of the political system and territorial integrity.

Chile has been well disposed to negotiate a territorial solution, provided that it will not interfere with the contiguity of its territory and

184

that Bolivia, in turn, will give a reliable territorial compensation. The only acceptable solution on these terms is the creation of a Bolivian territorial corridor which runs along the present Chilean–Peruvian boundary. Bolivia has agreed to this, but it does not accept territorial compensation to Chile. Now turning to Peru, when it has been asked by Chile about the subject, it has neither approved nor refused the idea; it has simply suggested some unacceptable territorial solutions for the Chilean government. With regard to this, the Bolivian Government has never expressed an opinion concerning the 1978 Peruvian proposals. The notion of 'territorial-state' is significant within the three states because it supports and fosters extreme nationalist approaches. Territorialism and nationalism with a territorial basis represent antagonistic approaches in any negotiation involving sovereign territories.

The notion of territorial replevy is common to members of Bolivian and Peruvian political classes, who forget that the validity of boundary jurisprudence is no longer based on historical rights, but on political transactions in force since the late-nineteenth century (1866, 1872, 1874, 1883, 1884). Chilean political society has shown that it is well disposed towards negotiations concerning territories and boundaries.

These considerations lead towards a global hypothesis which states that Bolivian maritime aspirations have not implemented the mechanism nor reached the adequate stage to achieve a territorial solution. If that happened, new aspirations and uncertainties would be bound to surface. Some Bolivians would perhaps insist on the notion of recovering the territory of Antofagasta, 'captive' under the Chilean political system. Peruvian intelligentsia could not accept that a sector of its southern coastal territories would finally revert to Bolivian hands.

ALTERNATIVE SOLUTIONS

Much legal, political and military evidence underlines the disadvantages for any Chilean territorial solution to Bolivian maritime aspirations, as it would directly affect Chilean land and maritime security, and would cause the breaking up of the balance introduced by M. Egaña (1825).

With regard to the jurisprudence produced by Chilean–Bolivian and Chilean–Peruvian agreements, the solution to Bolivia's maritime ambition should be located along the Chilean–Peruvian boundary. According to Lagos (1983), there are three alternatives for the achievement of such a solution. The first is that Peru could offer Bolivia a sovereign outlet to the sea through the territory of Tacna, which belongs to the former, close to the border with Chile.[1] A prior acceptance on the part of Chile would be necessary for its fulfilment. The second option is that Chile may offer a sovereign outlet to the sea through its territory of Arica, close to the border with Peru. Prior acceptance on the part of Peru is necessary for its fulfilment. The third option is that Chile and Peru may offer Bolivia a sovereign outlet to the sea through their territories between Tacna and Arica, located on either side of the boundary established in 1929.

Eventually, any of the three alternatives may create new and complex problems. Any of these options, if they could be realized, would meet the Bolivian territorial, nationalist and political aspirations; but they would go against the Peruvian and Chilean intelligentsia's way of thinking. This suggests alternative solutions for the Bolivian maritime aspiration, starting basically with the right for commercial traffic granted to Bolivia through Chilean territory and seaports. In 1987 Bolivian imports amounted to $776 million, whereas Chile imported products totalling $4,023 million. Bilateral trade, according to 1990 figures, greatly favoured the Chilean economy, since the cost of exports to Bolivia was $80 million, and Chilean imports from that country barely reached $46 million.

Within the framework of international law and international trade, as conceived at the end of the twentieth century, the interdependency of the states requires clearer and more practical solutions in order to meet the maritime aspirations by non-territorial routes.

Within the wide range of international relations, we observe three interesting economic systems: the economic cooperation and bilateral physical integration; common interests; and the economical integration at different trade levels, customs duties agreements, common market and economic confederation. The observation of resource availability and demand in the Chilean and Bolivian neighbouring territories shows that bilateral economic cooperation could be fruitful for both political systems. For instance, Bolivian hydraulic resources play an important role in the economic cooperation, whereas Chilean agricultural land can provide excellent opportunities for joint ventures such as irrigation projects.

The mechanisms both for economic cooperation and physical integration are incorporated in the 1904 Treaty, as well as the mechanisms for Bolivian trading through ports and railways from Arica to La Paz, and Antofagasta–Oruro–La Paz.

FINAL CONSIDERATIONS

From a legal standpoint, there are no territorial and/or boundary problems between Chile and Bolivia. However, the diplomatic relations – interrupted by Bolivia for a long time – have been overshadowed by Bolivian claims to Chilean coast. From a political and diplomatic perspective, there are no chances for an immediate solution. Probably, a satisfactory solution for Bolivia is to look for a more sound economic and commercial association with Chile, perhaps with Chile offering Bolivia various options to strengthen the initial economic cooperation and physical integration.[2]

NOTES

1 This solution has apparently become a reality in 1992, as Peru has ceded to Bolivia a Free Zone at the Port of Ilo.

2 On April 6 1993, the government of Chile and Bolivia, in the context and framework of the Latin American Association for Integration (ALADI), signed an Economic Complementation Agreement, which was put into effect on June 1 of that year.

BIBLIOGRAPHY

Abecia Baldivieso, Valentín (1979) *Las Relaciones Internationales en la Historia de Bolivia*, La Paz: Editorial 'Los Amigos del Libro'.

Anclarill, H. and Vilgre, M. (1977) 'La salida al mar de Bolivia', *Revista Argentina de Relaciones Internacionales*, vol. 8.

Aravena, R.N. (1987) *Un corredor territorial para Bolivia: ventajas y desventajas geopolíticas*, Santiago: Instituto de Ciencia Política, Universidad de Chile.

Baptista, J.M. (1978) *Tamayo y la reivindicación marítima*, La Paz: Ed. Casa Municipal de la Cultura.

Barros, B.L. (1922) *La cuestión del Pacífico y las nuevas orientaciones de Bolivia*, Santiago: Imprenta y Librería Artes y Letras.

Botelho, G.R. (1980) *El litoral boliviano: perspectiva histórica y geopolítica*, Buenos Aires: El Cid Editor.

Canelas, O.A. (1977) *Bolivia: mito y realidad de su enclaustramiento*, Lima: Editorial Tipografia y Offset Peruana.

Child, J. (1981) 'Pensamiento geopolítico y cuatro conflictos en Sudamérica', *Revista de Ciencia Política*, 1, 2.

Condarco, M.R. (1984) *Atlas histórico de Bolivia*, La Paz: Editorial Juventud.

Diaz, A.R. (1977) 'La respuesta chilena a Bolivia y el derecho internacional', *Panorama de la Política Mundial*, Santiago: Instituto de Estudios Internacionales, Universidad de Chile.

Encina, F. (1963) *Las relaciones entre Chile y Bolivia, 1841–1963*, Santiago: Nascimiento.

Escobari, J. (1975) *Historia diplomática de Bolivia*, La Paz: Ed. Universidad Boliviana.
Frontaura, A.M. (1968) *El Litoral de Bolivia*, La Paz: Editorial Burillo Ltd.
Lagos, C.G. (1981) *Límites y fronteras de Chile*, Santiago: Editorial Andrés Bello.
Meneses, R. (1943) *El imperativo geográfico en la mediterraneidad de Bolivia*, La Paz: Editorial Renacimiento.
Pereira, T. (1977) 'La consolidación territorial con los países limítrofes', *Cientocincuenta años de política chilena exterior*, Santiago: Instituto de Estudios Internacionales, Universidad de Chile.
Rios, G.C. (1963) *Chile y Bolivia definen sus fronteras, 1842–1904*, Santiago: Editorial Andres Bello.
Valencia, A. (1984) *Geopolítica en Bolivia*, La Paz: Juventud.

INDEX

agriculture, water needs of 35
Akwesasne, and Quebec–United
States boundary 14
alcohol restriction, and crossing
borders 6–7, 9
Alessandri, Arturo, President of Chile,
and relations with Bolivia 179
Alexander, E.P., and demarcation of
Nicaragua–Costa Rica border
100–1
de Alvarado, Pedro, Governor of
Guatemala and Chiapa 61–2
Alvarez, P., on Tate and Lyle in Belize
79
Alvarez-López, J.: on salts and
pesticides causing soil
deterioration 35; on water
pollution 36
Amazon region: biotechnology
development 145, 148; and
Brazil's frontier 141–6; disputes
over ownership of 121–4;
environmental concerns about
Brazilian development 144, 145,
148; and genetic engineering 145;
Grande Carajas Programme (PGC)
142, 143–4; process of Brazilian
development 142–5; reasons for
Brazilian development of 141–2
American Atlantic and Pacific
International Ship Company,
proposed canal across Nicaragua
91–2
Andes Mountains: 'Araucanization'

process 152–3, 155; as social
space 151; trading routes 151–2
Argentina: border with Chile see
Argentina–Chile border; economic
expansion and incorporation of
new lands 157–60;
industrialization phase 163; role in
world markets 155; war with
Paraguay 155
Argentina–Chile border 151–69; and
cattle rustling 153–5; crisis in
relations (1930s and 1940s)
163–7; customs duties 162; and
Desert Campaign (1879) 153–7;
enforcement of 152; smuggling
166; tensions 159, 161–2
armed forces: and authoritarianism in
South America 136; in Brazil 134,
136; modernization of weapons
139–40; opposition to 136
arms industry, Brazil 147–8
asymmetry in border regions: Central
America 52–3, 56;
Mexico–United States 19
Atacama Desert, and Bolivia–Chile
maritime dispute 170, 174–7
Aubertin, C. and Lena, P., on use of
frontier as symbol 138
Ayón, Tomás, President of Nicaragua
99–100

Baja California: climate 32–3;
disposal of waste water 36;
ecological interaction with

Printed and bound by CPI Group (UK) Ltd, Croydon, CR0 4YY

01/11/2024

01782616-0005